QUANTUM LOGIC

SYNTHESE LIBRARY

STUDIES IN EPISTEMOLOGY,

LOGIC, METHODOLOGY, AND PHILOSOPHY OF SCIENCE

Managing Editor:

JAAKKO HINTIKKA, *Academy of Finland and Stanford University*

Editors:

ROBERT S. COHEN, *Boston University*

DONALD DAVIDSON, *University of Chicago*

GABRIËL NUCHELMANS, *University of Leyden*

WESLEY C. SALMON, *University of Arizona*

VOLUME 126

QUANTUM LOGIC

by

PETER MITTELSTAEDT

University of Cologne, Germany

D. REIDEL PUBLISHING COMPANY

DORDRECHT : HOLLAND / BOSTON : U.S.A.

LONDON : ENGLAND

Library of Congress Cataloging in Publication Data

Mittelstaedt, Peter, 1929-
Quantum logic.

(Synthese library; v. 126)
Bibliography: p.
Includes index.
1. Quantum theory. 2. Logic, Symbolic and mathematical. I. Title.
QC174.17.M35M57 530.1'2 78-10433
ISBN 90-277-0925-4

Published by D. Reidel Publishing Company,
P.O. Box 17, Dordrecht, Holland

Sold and distributed in the U.S.A., Canada, and Mexico
by D. Reidel Publishing Company, Inc.
Lincoln Building, 160 Old Derby Street, Hingham, Mass. 02043, U.S.A.

Printed in The Netherlands

TO MECHTHILD

TABLE OF CONTENTS

INTRODUCTION

In 1936, G. Birkhoff and J. v. Neumann published an article with the title 'The logic of quantum mechanics'. In this paper, the authors demonstrated that in quantum mechanics the most simple observables which correspond to yes-no propositions about a quantum physical system constitute an algebraic structure, the most important properties of which are given by an orthocomplemented and quasimodular lattice L_q. Furthermore, this lattice of quantum mechanical propositions has, from a formal point of view, many similarities with a Boolean lattice L_B which is known to be the lattice of classical propositional logic. Therefore, one could conjecture that due to the algebraic structure of quantum mechanical observables a *logical calculus Q* of quantum mechanical propositions is established, which is slightly different from the calculus L of classical propositional logic but which is applicable to all quantum mechanical propositions (C.F. v. Weizsäcker, 1955). This calculus has sometimes been called '*quantum logic*'.

However, the statement that propositions about quantum physical systems are governed by the laws of quantum logic, which differ from ordinary classical logic and which are based on the empirically well-established quantum theory, is exposed to two serious objections:

(a) Logic is a theory which deals with those relationships between various propositions that are valid independent of the content of the respective propositions. Thus, the validity of logical relationships is not restricted to a special type of proposition, e.g. to propositions about classical physical systems.

(b) The laws of logic, though valid for all statements encountered in experience, do not derive their validity from experience. Instead, logic must be considered as a theory the statements of which can be justified exclusively by its inherent evidence and irrespective of all empirical knowledge.

These arguments become particularly apparent in the framework of the operational foundation of logic, which explains the 'truth' of a logical statement by the existence of a strategy of success in a dialog. In this approach, the laws of logic are completely determined by the possibilities of proving or disproving elementary and compound propositions within a well-defined proof procedure which can be represented by a dialog. Hence, the logical statements are 'true' independent of the special content of the actual propositions and for this reason they can neither be proved nor disproved by arguments which are based on empirical knowledge.

The controversial situation which thus arises can, however, be clarified if the dialogic justification of ordinary propositional logic is investigated in more detail. In this way, it turns out that in the framework of the dialogic method an assumption is always tacitly made which restricts the propositions considered to those of classical physics and mathematics. Technically, this means that within the dialogic proof procedure of compound propositions only the truth or falsity of the respective sub-propositions is tested but not their mutual commensurability. Therefore, one could try to search for an operational foundation of logic which is independent of the presupposition just mentioned. By this it is not meant that some emprical knowledge about quantum mechanical propositions should be incorporated into the foundation of logic. Conversely, we will rather eliminate that empirical supposition which is still contained in the operational foundation of ordinary logic – namely the assertion that all propositions are mutually commensurable. Technically, this generalization of the dialogic method can be performed by incorporating an additional testing procedure which – apart from the truth and falsity – examines the mutual commensurability of two propositions. In this way, the generalized dialog-game can be equally applied to propositions of classical physics and of quantum physics.

Starting from this generalized operational foundation of logic one arrives at a logical calculus Q_{eff}, which will be called the *calculus of effective quantum logic*, and which differs from the well-known calculus of effective (intuitionistic) logic inasmuch as some of the laws of this logic are valid only in a relaxed version. Furthermore, by a rather weak assumption concerning the measurability of commensurabilities, the '*tertium non datur*' can be justified for all finite compound propositions and the calculus Q_{eff} can be extended to the calculus Q of full quantum logic which incorporates the principle of excluded middle as a general law.

It can then be shown that this calculus Q of full quantum logic is, in fact, a model of the orthocomplemented and quasimodular lattice L_q which has been obtained previously from the algebraic structure of quantum mechanical observables. In this way, one obtains the following important results:

(a) The orthocomplemented quasimodular lattice L_q which follows from a formal analysis of quantum mechanics can be interpreted as a generalized propositional logic. This 'quantum logic' is universal in the sense that the validity of its laws is not restricted to a special type of proposition. Instead, the laws of quantum logic are equally valid for all propositions of classical physics and of quantum physics.

(b) Since the calculus Q of full quantum logic can be justified by theoretical reasons only and independent of any empirical knowledge, it follows that at least those structures of quantum mechanics which are summarized in the lattice L_q are non-empirical and may be considered as cognitions a priori.

In order to demonstrate these results, we proceed in the following way: In Chapter 1, we summarize some formal properties of quantum mechanics which are of special importance for the problem of quantum logic. It is shown that the yes-no propositions $A, B, \ldots$ about a physical system S correspond respectively to sub-spaces M_A, $M_B, \ldots$ of the Hilbert space $\mathcal{H}(S)$ which is associated with the system S as its state-space. On the other hand, these subspaces of a Hilbert space constitute a lattice, the most important properties of which are given by an orthocomplemented and quasimodular lattice L_q. Hence, quantum mechanical propositions form a lattice which has many properties in common with the Boolean lattice L_B of classical propositional logic and the operations of which have many similarities with the logical operations 'and', 'or' and 'not'.

The formal properties of the lattice L_q and its relations to a Boolean lattice L_B are investigated in Chapter 2. Firstly, we show that a commensurability relation can be defined in L_q such that mutually commensurable propositions form Boolean sublattices of L_q. Secondly, it is shown that the most important syntactical requirements for a logical calculus which can be formulated on the basis of L_B are also fulfilled by the lattice L_q. Hence, there are no formal arguments which exclude the possibility of an interpretation of L_q as a logical calculus. However, it is obvious that the formal similarities of L_q and L_B are by no means sufficient in order to justify a logical interpretation of the lattice L_q.

Therefore, for the present, we leave lattice theory and in Chapter 3 start from the very beginning by defining the first elements of a language of quantum physics. Apart from the elementary propositions, which can be proved or disproved by measurements, we introduce the logical connectives by means of dialogs. It already becomes apparent at this point that the language of quantum physics is more complex than the language of classical physics. In addition to elementary propositions, we use commensurability propositions as a means for testing the mutual commensurability of several propositions. In this way, it will be decided in every case whether two propositions are commensurable and thus the language considered can be applied to propositions of quantum physics as well as to propositions of classical physics.

After these preparations, the construction of a formal propositional logic is straightforward (Chapter 4). A compound proposition A is said to be formally true if there exists a strategy of success in a dialog irrespective of the truth or falsity of the subpropositions of A and irrespective of the mutual commensurabilities of these subpropositions. It is obvious what the generalisation of ordinary logic consists of: In ordinary logic, a proposition A is called formally true if there is a strategy of success irrespective of the truth and falsity of the subpropositions, whereas the mutual commensurabilities are always taken for granted. The totality of all formally true propositions is called the effective (intuitionistic) quantum propositional logic. It can be summarized in the calculus Q_{eff} of effective quantum logic, which turns out to be consistent and complete with respect to the formal quantum dialog-game.

In Chapter 5, we investigate the formal structure of the calculus Q_{eff} of effective quantum logic. It is shown that Q_{eff} is a model of a lattice L_{qi} which will be called the quasi-implicative lattice. It turns out that the lattice L_{qi} is a relaxation of the implicative lattice L_{i} as well as of the quasimodular lattice L_{q}. The lattice L_{qi} passes into the lattice L_{i} of ordinary effective logic if the commensurability of all propositions is presupposed. If all propositions are assumed to be either true or false (i.e. 'value-definite') the lattice L_{qi} passes into the orthocomplemented quasimodular lattice L_{q}.

The value-definiteness of propositions does not follow from the laws of logic but is an intrinsic property of the special type of proposition considered. The elementary propositions of quantum physics may be considered as value-definite. In Chapter 6, we show, within the calculus Q_{eff} by complete induction with respect to the logical operations, that the value-definiteness of elementary propositions implies the value-definiteness of all finite compound propositions (For this proof, a weak assumption must be made concerning the measurability of the commensurability propositions.) In this way, one arrives at the calculus Q of full quantum logic which incorporates the principle of excluded middle for all finite compound propositions. On the other hand, the calculus Q is a model of the orthocomplemented and quasimodular lattice L_q, and the logical interpretation of it is thus established.

The scheme summarizes this chain of reasoning and illustrates the interrelations between the various chapters of this book.

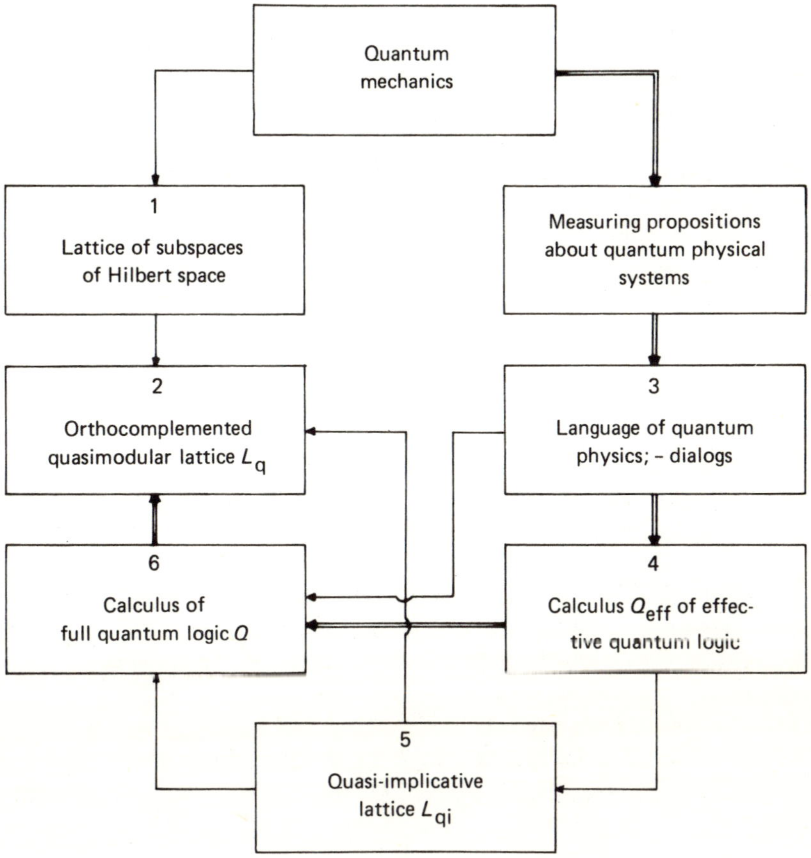

CHAPTER 1

THE HILBERT SPACE FORMULATION OF QUANTUM PHYSICS

In this chapter, we introduce the basic concepts of quantum theory. In Section 1.1, the state-space of a quantum physical system, the Hilbert space, is presented in axiomatic form and the concept of a closed linear manifold (subspace) is defined. In Section 1.2, we investigate the algebra of the subspaces of a Hilbert space and show that these subspaces form an *orthocomplemented quasimodular lattice*, which, moreover, has some additional properties. Closed linear manifolds are very closely related to *projection operators*, which are introduced in Section 1.3. On the other hand, projection operators represent observable quantities of the physical system. A physical system is characterized by its *state* and by its *properties*. These concepts will be defined in Section 1.4, and their relations to the elements and the subspaces of Hilbert space will be established.

In this introductory chapter, we summarize results which are for the most part well-established. Therefore, several theorems will be stated here without proofs and the reader will be referred to the literature.

1.1 THE HILBERT SPACE

1.1.1 *The Axioms of Hilbert Space*

The abstract Hilbert space $\mathcal{H}$ is a set of elements called vectors $f, g, \ldots$ which satisfy the following axioms:[1,2]

(I) *$\mathcal{H}$ is a linear vector space with complex coefficients*: i.e. for any two elements $f, g \in \mathcal{H}$ there exists an element $(f+g) \in \mathcal{H}$ and for every element $f \in \mathcal{H}$ and every complex number λ there exists an element $\lambda f \in \mathcal{H}$, such that the following relations are satisfied:

$$f+g=g+f,$$
$$(f+g)+h=f+(g+h),$$
$$\lambda(f+g)=\lambda f+\lambda g,$$

$$(\lambda + \mu)f = \lambda f + \mu f,$$
$$\lambda(\mu f) = (\lambda\mu)f, \qquad 1f = f.$$

Furthermore, there exists a unique zero element $\mathbf{0}$ such that for all $f \in \mathcal{H}$, we have

$$\mathbf{0} + f = f, \qquad 0 \cdot f = \mathbf{0}.$$

(II) *There exists a positive-definite scalar product in* $\mathcal{H}$: i.e. for any two elements $f, g \in \mathcal{H}$ there exists a complex number (f, g) which satisfies the following relations:

$$(f, g) = (g, f)^*,$$
$$(f, g + h) = (f, g) + (f, h),$$
$$\lambda(f, g) = (f, \lambda g), \qquad \text{for complex } \lambda,$$
$$(f, f) \equiv \|f\|^2 > 0, \qquad \text{unless } f = \mathbf{0}.$$

The (positive) expression $\|f\|$ is also called the norm of the vector f.

(III) $\mathcal{H}$ *is separable*: i.e. there exists a sequence $\{f_n\}$ of elements $f_n \in \mathcal{H}$ $(n = 1, 2, \ldots)$ with the property that for any $f \in \mathcal{H}$ and $\epsilon > 0$ there exists at least one element $f_n \in \{f_i\}$ such that

$$\|f - f_n\| < \epsilon.$$

The sequence f_n is also said to be dense in $\mathcal{H}$.

(IV) $\mathcal{H}$ *is complete*; i.e. any sequence f_n of elements $f_i \in \mathcal{H}$, with the property

$$\lim_{n,m\to\infty} \|f_n - f_m\| = 0,$$

has a uniquely defined limit $f \in \mathcal{H}$ such that

$$\lim_{n\to\infty} \|f - f_n\| = 0.$$

In addition to the Axioms I–IV of Hilbert space, some useful concepts will be defined and briefly commented upon. A finite or infinite sequence $\{f_n\}$ of vectors $f_n \in \mathcal{H}$ is called *linearly independent* if a relation of the kind $\sum_n \lambda_n f_n = 0$ implies $\lambda_n = 0$ for all n. The maximal number $d = d(\mathcal{H})$ of linearly independent vectors is called the *dimension* $d(\mathcal{H})$ of $\mathcal{H}$. The Axioms I–IV of Hilbert space do not specify whether $\mathcal{H}$ is a finite- or an infinite-dimensional space, since for the application to quantum theory it is more convenient to have a definition which covers both cases. If the dimension $d = d(\mathcal{H})$ is finite, the Axioms I–IV are not independent, since, in that case, III and IV are consequences of the others. If $d(\mathcal{H}) = \infty$, which corresponds to the usual definition of $\mathcal{H}$, the Axioms I–IV are independent.

A Hilbert space is not only a vector space but also a topological space. A notion of convergence can be defined either by means of the norm or the scalar product. A sequence $\{f_n\}$ is said to converge in the norm to the vector f, if

$$\lim_{n\to\infty} \|f_n - f\| = 0.$$

In this case, we have *strong convergence.* Using the scalar product, a sequence $\{f_n\}$ is said to *converge weakly* to f, if for every vector $g \in \mathcal{H}$ we have

$$\lim_{n\to\infty} (f_n, g) = (f, g).$$

In finite dimensional Hilbert spaces the two concepts of convergence are equivalent, but not in the infinite dimensional case.

Two vectors $f, g \in \mathcal{H}$ with $f \neq \mathbf{0}$ and $g \neq \mathbf{0}$, which satisfy the relation $(f, g) = 0$, are said to be *orthogonal.* A sequence $\{\varphi_\nu\}$ of vectors $\varphi_\nu \in \mathcal{H}$, $\varphi_\nu \neq \mathbf{0}$ is called *orthonormal,* if for all pairs of elements $\varphi_\nu, \varphi_\mu \in \{\varphi_\lambda\}$ the equation $(\varphi_\nu, \varphi_\mu) = \delta_{\nu\mu}$ holds. Furthermore, an orthonormal sequence $\{\varphi_\nu\}$ is called *complete* if for any vector $f \in \mathcal{H}$ the relation

$$\sum_{\nu=1}^{\infty} |(\varphi_\nu, f)|^2 = \|f\|^2$$

is fulfilled. In this case, the partial sums $f^{(n)} = \Sigma_{\nu=1}^{n} (\varphi_\nu, f)\varphi_\nu$ converge strongly to f, i.e. we have

$$\lim_{n\to\infty} \|f^{(n)} - f\| = 0.$$

Therefore, we may also write $f = \Sigma_{\nu=1}^{\infty} (\varphi_\nu, f)\varphi_\nu$. Such a sequence is called a *basis* or a *coordinate system* in Hilbert space. The existence of an orthonormal and complete sequence $\{\varphi_\nu\}$ is guaranteed by Axiom III.

1.1.2 *Linear Manifolds and Subspaces*

Let $\mathcal{M}$ be a subset $\mathcal{M} \leqslant \mathcal{H}$ of $\mathcal{H}$. A subset is called a *linear manifold* $\mathcal{M}^{(L)}$, if for all elements $f, g \in \mathcal{M}^{(L)}$ we have

$$f \in \mathcal{M}^{(L)} \curvearrowright \lambda f \in \mathcal{M}^{(L)},$$
$$f, g \in \mathcal{M}^{(L)} \curvearrowright (f + g) \in \mathcal{M}^{(L)},$$

for complex numbers λ. For a linear manifold, the Axioms I, II and III for a Hilbert space are fulfilled, but, in general, Axiom IV is not.

A linear manifold is called a *closed linear manifold* M if for any sequence $\{f_n\}$ with $f_n \in M$, which has a limit vector f, this limit vector also belongs to M; i.e. $f \in M$. Closed linear manifolds will also be called *subspaces* and will be denoted by $M_A, M_B, \ldots$. A subspace of a Hilbert space is itself a Hilbert space, i.e. Axioms I, II, III and IV are always fulfilled for a closed linear manifold.[3,4,5]

On the set $\mathscr{L}$ of closed linear manifolds we will now define some operations and one relation which lead to an interesting algebraic structure:

(α) A 2-place relation $R \subseteq \mathscr{L} \times \mathscr{L}$ is given by the set theoretical inclusion '$\subseteq$' between two subsets M_A and M_B. It is obvious that this relation fulfills the conditions

$$M_A \subseteq M_A \tag{1.1}$$
$$M_A \subseteq M_B,\ M_B \subseteq M_C \curvearrowright M_A \subseteq M_C.$$

Furthermore, the set-theoretical equivalence '=' is related to the relation R by

$$M_A \subseteq M_B,\ M_B \subseteq M_A \curvearrowright M_A = M_B.$$

(β) A 1-place operation $\Theta_\perp : \mathscr{L} \to \mathscr{L}$ can be defined as follows. Let M_A be a subspace. The set $\mathscr{M}_\perp(M_A)$ of elements, which are orthogonal to all vectors of M_A, i.e.

$$\mathscr{M}_\perp(M_A) := \{f : (f, g) = 0 \quad \text{for all } g \in M_A\},$$

is also a subspace and will be denoted by $\perp M_A$. The subspace $\perp M_A$ is said to be *completely orthogonal* to M_A. From the definition it is easily seen that the equation

$$M_A = \perp(\perp M_A) \tag{1.2}$$

is valid. Furthermore, if M_A and M_B are subspaces such that $M_A \subseteq M_B$ it follows that

$$M_A \subseteq M_B \curvearrowright \perp M_B \curvearrowright \perp M_A \tag{1.3}$$

is valid.

(γ) The 2-place operation $\Theta_\cap : \mathscr{L} \times \mathscr{L} \to \mathscr{L}$ will be defined in the following way. If M_A and M_B are subspaces, the set $\mathscr{M}_\cap(M_A, M_B)$ of elements

$$\mathcal{M}_\cap(M_A, M_B) := \{f : f \in M_A \quad \text{and} \quad f \in M_B\},$$

which corresponds to the intersection of the sets M_A and M_B, is also a subspace and will be denoted by $M_A \cap M_B$. Obviously, we have

$$M_A \cap M_B \subseteq M_A, \qquad M_A \cap M_B \subseteq M_B. \tag{1.4}$$

Furthermore, it is easily seen that the intersection $M_A \cap M_B$ is the largest subspace contained in M_A and M_B (infimum); i.e.

$$M_C \subseteq M_A, M_C \subseteq M_B \curvearrowright M_C \subseteq M_A \cap M_B. \tag{1.5}$$

If $\{M_i\}$ is a finite family of subsets $(i = 1, 2, \ldots)$, the intersection of these subspaces will be denoted by

$$\bigcap_i M_i = M_1 \cap M_2 \cap \ldots$$

(δ) The 2-place operation $\Theta_\cup : \mathcal{L} \times \mathcal{L} \to \mathcal{L}$ can be defined as follows. If M_A and M_B are subspaces, the set $M_\cup(M_A, M_B)$ consists of all elements $h = \alpha f + \beta g$ with $f \in M_A$, $g \in M_B$ for complex numbers α, β and of the limit vectors. Therefore, $\mathcal{M}_\cup(M_A, M_B)$ is a closed linear manifold which will be denoted here by $M_A \cup M_B$. The subspace $M_A \cup M_B$ is said to be spanned by the elements $f \in M_A$ and $g \in M_B$. Obviously we have

$$M_A \subseteq M_A \cup M_B, \qquad M_B \subseteq M_A \cup M_B. \tag{1.6}$$

Furthermore, one finds that $M_A \cup M_B$ is the smallest subset which contains M_A and M_B (supremum); i.e.

$$M_A \subseteq M_C, M_B \subseteq M_C \curvearrowright M_A \cup M_B \subseteq M_C \tag{1.7}$$

is valid for arbitrary M_A, M_B, M_C. If $\{M_i\}$ is a finite family of subspaces $(i = 1, 2, \ldots)$, for the subspace spanned by the M_i we use the notation

$$\bigcup_i M_i = M_1 \cup M_2 \cup \ldots .$$

An important consequence of these definitions is the following decomposition property of an arbitrary vector $f \in \mathcal{H}$ with respect to a subspace M_A. To every subspace M_A there belongs a *unique* decomposition of f such that $f = f_A + f_{\perp A}$, with $f_A \in M_A$ and $f_{\perp A} \in \perp M_A$. The

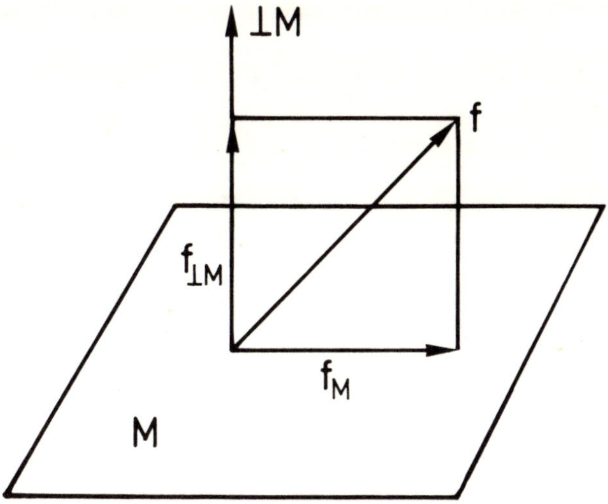

Fig. 1.1. The decomposition of a vector f with respect to the subspace M.

geometrical meaning of this theorem is obvious (Fig. 1.1). On the other hand, the two subspaces M_A and $\perp M_A$ span the entire space; i.e. $M_A \cup \perp M_A = \mathscr{H}$. Thus, the decomposition can be generalized to more than one subspace. Let $\{M_\nu\}$ be a (finite or infinite) sequence of mutually orthogonal subspaces such that $\perp(M_1 \cup M_2 \cup \ldots) = \mathbf{0}$; i.e., the subspaces M_ν span the whole Hilbert space $\mathscr{H}$. An arbitrary $f \in \mathscr{H}$ can then uniquely be decomposed into a sum $f = \sum_{\nu=1}^{\infty} f_\nu$, with $f_\nu \in M_\nu$, of mutually orthogonal vectors f_ν. (The infinite sum has to be interpreted in the sense of strong convergence.)

1.2 THE LATTICE OF SUBSPACES OF HILBERT SPACE

We shall now consider the algebraic structure which is generated on the set $\mathscr{L}$ of closed linear manifolds by

the 2-place relation	$R \subseteq \mathscr{L} \times \mathscr{L}$,
the 1-place operation	$\Theta_\perp : \mathscr{L} \to \mathscr{L}$,
the 2-place operations	$\Theta_\cap : \mathscr{L} \times \mathscr{L} \to \mathscr{L}$,
	$\Theta_\cup : \mathscr{L} \times \mathscr{L} \to \mathscr{L}$.

I.e. we shall investigate the structure of the algebra $L_H = \langle \mathscr{L}; \subseteq, \perp, \cap, \cup \rangle$.[5,6,7]

From the definition of the relation R, according to (1.1), it follows that

$L_H(1.1)$ $\quad M_A \subseteq M_A$,

$L_H(1.2)$ $\quad M_A \subseteq M_B,\ M_B \subseteq M_C \curvearrowright M_A \subseteq M_C$,

$L_H(1.3)$ $\quad M_A \subseteq M_B,\ M_B \subseteq M_A \curvearrowright M_A \curvearrowright M_A = M_B$.

I.e. with respect to the relation '$\subseteq$', $\mathscr{L}$ is a partially-ordered set. The condition $L_H(1.3)$ connects the partial-ordering relation '$\subseteq$' with the equivalence relation '='. Furthermore, from the definitions of the operations $\Theta_\cap$ and $\Theta_\cup$ we obtain, according to (1.4), (1.5), (1.6) and (1.7),

$L_H(2.1)$ $\quad M_A \cap M_B \subseteq M_A$,

$L_H(2.2)$ $\quad M_A \cap M_B \subseteq M_B$,

$L_H(2.3)$ $\quad M_C \subseteq M_A,\ M_C \subseteq M_B \curvearrowright M_C \subseteq M_A \cap M_B$,

$L_H(3.1)$ $\quad M_A \subseteq M_A \cup M_B$,

$L_H(3.2)$ $\quad M_B \subseteq M_A \cup M_B$,

$L_H(3.3)$ $\quad M_A \subseteq M_C,\ M_B \subseteq M_C \curvearrowright M_A \cup M_B \subseteq M_C$.

These relations $L_H(2.1)$–$L_H(3.3)$ state that for any two elements $M_A, M_B \in \mathscr{L}$ there exists with respect to the partial-ordering relation '$\subseteq$' a greatest lower bound $M_A \cap M_B$ (infimum) and a least upper bound $M_A \cup M_B$ (supremum). A partially-ordered set $\mathscr{L}$ which has this property is called a lattice $L = L(\mathscr{L})$. This property can be generalized to countably infinite sets of subspaces; i.e. for any countable set $\{M_\nu\}$ of subspaces there exists a greatest lower bound $\bigcap_{\nu=1}^{\infty} M_\nu$ and a least upper bound $\bigcup_{\nu=1}^{\infty} M_\nu$, both of which are subspaces of $\mathscr{H}$. A lattice with this property is called *σ-complete*. The *σ-completeness* of the lattice L_H is a consequence of Axiom IV which postulates the completeness of the Hilbert space.

Furthermore, the lattice of subspaces considered here has a zero element $M_0 = \mathbf{0}$ and a unit element $M_I = \mathscr{H}$, which satisfy the relations

$L_H(4.0)$ $\quad M_0 \subseteq M_A,\ M_A \subseteq M_I \quad$ for all M_A.

The operation $\Theta_\perp$, which maps a subspace $M_A \in \mathscr{L}$ to the completely orthogonal subspace $\perp M_A \in \mathcal{L}$ is an automorphism which satisfies the relations (cf. (1.2) and (1.3))

$L_H(4.1)$ $M_A \cap \perp M_A \subseteq M_0$,

$L_H(4.2)$ $M_I \subseteq M_A \cup \perp M_A$,

$L_H(4.3)$ $M_A = \perp(\perp M_A)$,

$L_H(4.4)$ $M_A \subseteq M_B \curvearrowright \perp M_B \subseteq \perp M_A$.

If, in a lattice L with elements M_0 and M_I, an *automorphism* $M_A \to \perp M_A$ is defined which fulfills the conditions $L_H(4.1)$–$L_H(4.4)$, this lattice is said to be *orthocomplemented*. The element $\perp M_A$ is called the *orthocomplement* of M_A. It should be mentioned that, in general, the element $\perp M_A$ is not uniquely defined by the lattice axioms $L_H(4.1)$–$L_H(4.4)$ but only by the automorphism. In the present case, this automorphism is the mapping to the completely orthogonal subspace.

The proofs of the lattice properties $L_H(1.1)$–$L_H(4.4)$ are very simple and will not be presented here. There are, however, more complicated properties of the lattice L_H, which should be discussed in greater detail. The lattice of subspaces L_H is not distributive; i.e. the relations

$$M_A \cap (M_B \cup M_C) = (M_A \cap M_B) \cup (M_A \cap M_C), \quad (1.8)$$

$$M_A \cup (M_B \cap M_C) = (M_A \cup M_B) \cap (M_A \cup M_C), \quad (1.8\dagger)$$

are not generally valid. However, some relaxations of the distributive law can be proved in L_H.

(1.9) THEOREM: (Modularity.) If $\mathcal{H}$ is a finite-dimensional space, then the lattice L_H is modular; i.e. the law

$$M_B \subseteq M_A \curvearrowright M_A \cap (M_B \cup M_C) \subseteq M_B \cup (M_A \cap M_C)$$

is valid.

Remark: Modularity is, in fact, a relaxation of distributivity since the modular law can also be written in the equivalent form

$$M_B \subseteq M_A \curvearrowright M_A \cap (M_B \cup M_C) = (M_A \cap M_B) \cup (M_A \cap M_C), \quad (1.10)$$

which shows that the distributive law (1.8†) is stated here only under the additional assumption $M_B \subseteq M_A$. The modular law is self-dual; i.e.

it follows from (1.10) that

$$M_B \subseteq M_A \curvearrowright M_B \cup (M_A \cap M_C) = (M_B \cup M_A) \cap (M_B \cup M_C) \tag{1.10†}$$

and vice versa. Therefore, only one form of the modular law is required.

Proof: If $f \in M_A \cap (M_B \cup M_C)$, the vector f can be decomposed into $f = f_B + f_C$ with $f_B \in M_B$, $f_C \in M_C$. Furthermore, $f = f_A$ with $f_A \in M_A$. Therefore, $f_C = f - f_B \in M_A$, since $M_B \subseteq M_A$ by assumption. From $f_C \in M_C$ we then obtain $f_C \in M_A \cap M_C$ and therefore $f = f_B + f_C \in M_B \cup (M_A \cap M_C)$, completing the proof.]

The modularity of L_H is restricted to finite-dimensional Hilbert spaces. This can easily be demonstrated by means of a counter-example (cf. ref. 7).

Let φ_n be an orthonormal basis in the Hilbert space $\mathscr{H}$ and β_n, γ_n elements defined by

$$\beta_n = \varphi_{2n} + a^{-n}\varphi_1 + a^{-2n}\varphi_{2n+1},$$

$$\gamma_n = \varphi_{2n},$$

with $(n = 1, 2, \ldots)$ and $a > 1$ real. We then consider the subspaces

$$M_B = [\beta_n], \qquad M_C = [\gamma_n], \qquad M_A = [\beta_n, \varphi_1]$$

spanned by the elements $\{\beta_n\}$, $\{\gamma_n\}$ and $\{\beta_n, \varphi_1\}$, respectively. (We write $[\alpha_n] = [\alpha_1, \alpha_2 \ldots]$ for the closed linear manifold spanned by the vectors $\{\alpha_n\}$.)

For these subspaces, we have $M_B \subseteq M_A$; i.e. the premise of the modular law is satisfied. Furthermore, we have $M_B \cap M_C = \mathbf{0}$ and $M_C \cap M_A = \mathbf{0}$ for the finite- as well as for the infinite-dimensional case. However, for a finite-dimensional Hilbert space we obtain $M_B \cup M_C \subsetneq M_A \cup M_C$, whereas for an infinite number of dimensions one obtains $M_B \cup M_C = M_A \cup M_C$. In fact, in the finite-dimensional case, φ_1 is an element of $M_A \cup M_C$ but not of $M_B \cup M_C$. In the infinite-dimensional case,

$$\xi_n = a^n\beta_n - a^n\gamma_n = \varphi_1 + a^{-n}\varphi_{2n+1} \in M_B \cup M_C$$

are elements of the closed linear manifold $M_B \cup M_C$ and, consequently, the limit vector

$$\xi := \lim_{n \to \infty} \xi_n = \varphi_1 \in M_B \cup M_C$$

which turns out to be equal to φ_1, is also such. Thus, in the infinite-dimensional case, we obtain $M_B \cup M_C = M_A \cup M_C$.

According to the modular law, the premise $M_B \subseteq M_A$ implies the conclusion $M_A \cap (M_B \cup M_C) \subseteq M_B \cup (M_A \cap M_C)$. However, in the example considered here, we obtain for the left-hand side $M_A \cap (M_B \cup M_C) = M_A \cap (M_A \cup M_C) = M_A$ and $M_B \cup (M_A \cap M_C) = M_B$ for the right-hand side; i.e. $M_A \subseteq M_B$. Therefore, if we have in the premise $M_B \subset M_A$ (i.e. $M_B \neq M_A$) then the conclusion of the modular law is false, and we arrive at the following result:

(1.11) THEOREM: If $\mathscr{H}$ is an infinite-dimensional space, then the lattice L_{H} is *not modular.*

In the proof of Theorem (1.9) the finite dimensionality has been used in decomposing the vector $f \in M_B \cup M_C$ into $f = f_B + f_C$. Obviously this is not possible for the limit vector $\xi = \lim_{n\to\infty} \xi_n$. This decisive step in the proof can be avoided if the modularity is weakened by an additional premise.

(1.12) THEOREM: (Quasimodularity.) The lattice L_{H} of subspaces of Hilbert space is *quasimodular*; i.e. the law

$L_{\mathrm{H}}(5)$ $$M_B \subseteq M_A,\ M_C \subseteq \perp M_A \curvearrowright M_A \cap (M_B \cup M_C) \subseteq M_B \cup (M_A \cap M_C)$$

is valid.

Remark: Quasimodularity is a relaxation of modularity since the conclusion $M_A \cap (M_B \cup M_C) \subseteq M_B \cup (M_A \cap M_C)$ is stated here under the additional premise $M_C \subseteq \perp M_A$.

Proof: From the premises, we obtain for vectors f_B and f_C: $f_B \in M_A$, $f_C \in \perp M_A$. If $f \in M_A \cap (M_B \cup M_C)$, we have $f \in M_A$ and $f \in M_B \cup M_C$. Therefore, f has no component in M_C, since f_C would be an element of $\perp M_A$. Consequently, $f \in M_B$ and also $f \in M_B \cup (M_A \cap M_C)$.⟧

Remark: In this proof, it is no longer necessary to decompose f into $f = f_B + f_C$ since, according to the new premise, namely $M_C \subseteq \perp M_A$, the vector f has no component in M_C.

Summarizing the results we have obtained up to now, the algebra $L_{\mathrm{H}} = \langle \mathscr{L};\ \subseteq,\ \perp,\ \cap,\ \cup \rangle$ of subspaces of Hilbert space is given by a σ-complete, orthocomplemented and quasimodular lattice, which is represented by the axioms $L_{\mathrm{H}}(1.1)$–$L_{\mathrm{H}}(5)$.

In the following chapters, only those properties of the lattice L_{H}

which follow from the axioms $L_H(1.1)$–$L_H(5)$ will be important. There are, however, two additional properties of L_H – the 'atomicity' and the 'covering law' – which are interesting in many respects and which should be mentioned here briefly for the sake of completeness.

(1.13) DEFINITION: If L is a lattice and a, b are elements with $a \subseteq b$ and $a \neq b$, the b is said to *cover* a if $a \subseteq x \subseteq b$ implies $a = x$ or $x = b$. An element α which covers the zero element $\mathbf{0}$ is called an *atom*. If for any element x of a lattice L there exists an atom α with $\alpha \subseteq x$ this lattice is said to be *atomic*.

(1.14) THEOREM: (Atomicity.) The lattice L_H of subspaces of Hilbert space is *atomic*.

Proof: Let $M^x = [f_\lambda^x] \in L_H$ be the subspace spanned by the vectors f_λ^x and $f = \Sigma_\lambda c_\lambda f_\lambda^x$ a one-dimensional element $f \in M^x$. Then the subspace $M^f = [f]$ is an *atom*. In fact, from $\mathbf{0} \subseteq M \subseteq M^f$ it follows that $g \in M$ implies $g \in M^f$, i.e. $g = f \cdot c$. If $c = 0$, we have $M = \mathbf{0}$ and if $c \neq 0$, it follows $M = M^f$.]

(1.15) THEOREM: (Covering law.) If $M^\alpha \in L_H$ is an atom, then for every element $M^a \in L_H$ the element $M^a \cup M^\alpha$ *covers* M^a.

Proof: Let $M^a = [f_\lambda^a]$ and $M^\alpha = [f]$. Furthermore, $M^a \cup M^\alpha = [f_\lambda^a, f]$. If for a subspace M^z we have $M^a \subseteq M^z \subseteq M^a \cup M^\alpha$ an element $g \in M^z \subseteq M^a \cup M^\alpha$ can be decomposed into $g = \Sigma_\lambda f_\lambda^a c_\lambda + f \cdot c$. If $c = 0$ for all $g \in M^z$, we get $M^z = M^a$. If for some $g \in M^z$ $c \neq 0$, it follows $M^z = M^a \cup M^\alpha$.]

1.3 PROJECTION OPERATORS

1.3.1 *Bounded Linear Operators in Hilbert Space*

(1.16) DEFINITION: A bounded linear operator is a function $T: D_T \to \Delta_T$, D_T, $\Delta_T \in \mathcal{H}$ with linear manifold D_T as *domain* and a subset Δ_T in $\mathcal{H}$ as *range* such that

$$T(f+g) = Tf + Tg, \quad f, g \in D_T,$$
$$T(\lambda f) = \lambda T(f), \quad f \in D_T, \ \lambda \text{ complex},$$
$$\|Tf\| \leq M\|f\|, \quad 0 \leq M < \infty.$$

It is a consequence of this definition that the range Δ_T is also a linear manifold. The greatest lower bound of the numbers M is called the norm $\|T\|$ of T. For bounded linear operators, the existence of an $M < \infty$ is postulated.

If from $f_1 \neq f_2$ it follows that $Tf_1 \neq Tf_2$, then there exists a linear operator T^{-1}, the *inverse* of T with the domain $D_{T^{-1}} = \Delta_T$ and the range $\Delta_{T^{-1}} = D_T$. This operator has the property

$$T^{-1}Tf = f, \quad f \in D_T,$$
$$TT^{-1}g = g, \quad g \in T.$$

The *sum* of two linear operators T_1 and T_2 is defined by

$$(T_1 + T_2)f = T_1f + T_2f,$$

and the *product* $T_1 \cdot T_2$ by

$$T_1T_2f = T_1(T_2f).$$

Furthermore, the multiplication with a complex λ is defined by

$$(\lambda T)f = \lambda Tf.$$

It follows from these definitions that if T_1 and T_2 are bounded linear operators, then $T_1 + T_2$, T_1T_2 and λT_1 are also bounded.

(1.17) DEFINITION: Two operators T and T^* are called *adjoint* to one another if $(T^*f, g) = (f, Tg)$. An operator with $T = T^*$ is called *self-adjoint.*

From this definition, the following relations can easily be shown to hold

$$(T_1T_2)^* = T_2^*T_1^*,$$
$$(\lambda T)^* = \lambda^*T^*, \quad \lambda \text{ complex},$$
$$(T_1 + T_2)^* = T_1^* + T_2^*,$$
$$(T^*)^* = T.$$

The bounded linear and self-adjoint operators play an important role in quantum theory. They correspond to physical observables and their spectra are related to possible measuring values. Without discussing here this connection in more detail, we first introduce the simplest operators of this kind, the projection operators.

1.3.2 *Projection Operators*

According to the decomposition theorem mentioned above, to every subspace $M_A \subseteq \mathcal{H}$ and to every $f \in \mathcal{H}$ there belongs a unique decomposition such that $f = f_A + f_{\perp A}$ with $f_A \in M_A$, $f_{\perp A} \in \perp M_A$. The vector f_A is called the projection of f onto M_A. The function

$$P_A: \mathcal{H} \to M_A, \qquad D_{P_A} = \mathcal{H}, \qquad \Delta_{P_A} = M_A$$

with $P_A f = f_A$ defines an operator P_A, the *projection operator*, which maps a vector $f \in \mathcal{H}$ to its projection f_A in M_A. The properties of a projection operator are given by the following theorem.

(1.18) THEOREM: A projection operator P_A which projects onto the subspace M_A is an operator with the properties:

(a) linear and bounded;
(b) self-adjoint ($P_A = P_A^*$);
(c) idempotent ($P_A^2 = P_A$);
(d) $D_{P_A} = \mathcal{H}$, $\Delta_{P_A} = M_A$;
(e) $P_A f = f$ iff $f \in M_A$,
$P_A f = 0$ iff $f \in \perp M_A$.

Proof: The *linearity* of P_A is obvious. If $f = f_A + f_{\perp A}$ with $f_A \in M_A$ and $f_{\perp A} \in \perp M_A$, we obtain

$$\|P_A f\|^2 = \|f_A\|^2 \leq \|f_A\|^2 + \|f_{\perp A}\|^2 = \|f\|^2,$$

so that P_A is *bounded* and $\|P_A\| \leq 1$. Furthermore, if $f = f_A + f_{\perp A}$ and $g = g_A + g_{\perp A}$ are vectors with $f_A, g_A \in M_A$ and $f_{\perp A}, g_{\perp A} \in \perp M_A$, we obtain

$$(P_A f, g) = (f_A, g) = (f_A, g_A) = (f, g_A) = (f, P_A g),$$

so that P_A is *self-adjoint*. Moreover, from

$$P_A(P_A f) = P_A f_A = f_A = P_A f$$

it follows that $P_A^2 = P_A$; i.e. P_A is *idempotent*. According to the definition of P_A, it is defined everywhere ($D_{P_A} = \mathcal{H}$) and the elements $P_A f = f_A$ belong to M_A; i.e. $\Delta_{P_A} = M_A$. If for a vector f the equation $P_A f = f$ holds, it follows $f = f_A$ and $f \in M_A$. Conversely, if $f \in M_A$, we have $P_A f = f$. If $P_A f = \mathbf{0}$, it follows from the decomposition $f = f_A + f_{\perp A}$ that $f_A = \mathbf{0}$; i.e., $f \in \perp M_A$. Conversely, if $f \in \perp M_A$, we obtain $P_A f = P_A f_{\perp A} = \mathbf{0}$.]]

The properties of a projection operator mentioned in (1.18) are also sufficient in order to characterize an operator as a projection operator. This can be done according to the following theorem:

(1.19) THEOREM: A bounded linear operator P defined everywhere ($D_P = \mathcal{H}$), which is self-adjoint ($P = P^*$) and idempotent ($P = P^2$), is a projection operator P_M, which maps on some subspace M uniquely defined by P_M.

Proof: Let M be the subspace spanned by all elements Pf. The vector $g - Pg$ is orthogonal to every Pf since

$$(Pf, g - Pg) = (Pf, g) - (Pf, Pg) = (Pf, g) - (P^2 f, g) = 0.$$

Furthermore, the elements which are orthogonal to $g - Pg$ span a subspace which, together with the elements Pf, also contains M. Hence $g - Pg$ is contained in $\perp M$ and the decomposition of g with respect to M is given by $g = Pg + (g - P \cdot g)$. Therefore, we obtain $P_M g = Pg$, where P_M is the projection operator which projects on M. Since g is arbitrary, we finally obtain $P_M = P$.]]

Summarizing the results of Theorems (1.18) and (1.19), we find that every projection operator P defines uniquely a subspace M and vice versa. Furthermore, the equations $Pf = f$ and $Pf = \mathbf{0}$ hold if and only if we have $f \in M$ and $f \in \perp M$, respectively.

1.3.3 *Projection Operators and Subspaces*

The one-to-one correspondence between projection operators and subspaces can be used in order to express the relation R between subspaces in terms of projection operators and to define the projection operators $P_{A \cap B}$, $P_{A \cup B}$ and $P_{\perp A}$ which correspond to the subspaces $M_{A \cap B}$, $M_{A \cup B}$ and $M_{\perp A}$, respectively. In this way, the entire algebraic structure of subspaces can be transferred to projection operators.

The partial-ordering relation '$\subseteq$' between subspaces can be expressed by means of projection operators due to the following theorem:

(1.20) THEOREM: If P_A, P_B are projection operators and M_A, M_B the corresponding subspaces, then the following three conditions are equivalent:

(a) $\|P_A f\| \leq \|P_B f\|$,

(b) $M_A \subseteq M_B$,

(c) $P_A = P_A P_B = P_B P_A$.

Proof: If $\|P_A f\| \leq \|P_B f\|$ for all f and if, in particular, $f \in M_A$, then we obtain $\|f\| = \|P_A f\| \leq \|P_B f\| \leq \|f\|$, since $\|P_B\| \leq 1$. Hence, we obtain $\|P_B f\| = \|f\|$ and $P_B f = f$; i.e. $f \in M_B$. Therefore, we have $M_A \subseteq M_B$. If $M_A \subseteq M_B$ then $P_A f \in M_B$ and $P_B P_A f = P_A f$ for every f; i.e. $P_B P_A = P_A$. Since $P_A = P_A^*$, it follows from this equation that $P_A P_B = P_B P_A = P_A$. If $P_A P_B = P_A$, we obtain for all f $(P_A f, f) = \|P_A f\|^2 = \|P_A P_B f\|^2 \leq \|P_B f\|^2$.]

In order to construct the projection operators which correspond to the operations $\cap$, $\cup$ and $\perp$, we begin with the projection operator $P_{\perp A}$ which projects to the subspace $M_{\perp A}$ which is completely orthogonal to M_A.

(1.21) THEOREM: If P is a projection operator with the range M, then $1 - P$ is the projection operator which projects on the subspace $\perp M$.

Proof: Since $(1-P)^* = 1 - P$ and $(1-P)^2 = 1 - P - P + P^2 = 1 - P$, the operator $(1-P)$ is a projection operator. Furthermore, for every f we have $(1-P)f = f - f_M = (f_M + f_{\perp M}) - f_M = f_{\perp M} \in \perp M$.]

For the construction of the projection operators $P_{A \cap B}$ and $P_{A \cup B}$ which correspond to the subspaces $M_A \cap M_B$ and $M_A \cup M_B$, respectively, it is important to note that the product $P_A \cdot P_B$ and the difference $P_A - P_B$ of two projection operators in general are not projection operators. This is the case if and only if $[P_A, P_B] = 0$ and if $P_B = P_B P_A$, respectively. In the general situation, the projection operators $P_{A \cap B}$ and $P_{A \cup B}$ are established in accordance with the following theorems.

(1.22) THEOREM: The projection operator $P_{A \cap B}$ with the range $M_A \cap M_B$ can be expressed by the projection operators P_A and P_B with the range M_A and M_B, respectively, due to the relation

$$P_{A \cap B} = \lim_{n \to \infty} (P_A P_B)^n.$$

Proof: The subspace $M_A \cap M_B$ is the infimum of M_A and M_B. Therefore, by virtue of Theorem (1.20), it follows that $P_{A \cap B}$ fulfills the conditions

(a) $P_{A\cap B} = P_A P_{A\cap B} = P_B P_{A\cap B}$,

(b) $P_C = P_C P_A = P_C P_B$ implies $P_C = P_C P_{A\cap B}$,

which determine $P_{A\cap B}$ uniquely. In fact, since $P := \lim_{n\to\infty} (P_A P_B)^n = P_A \cdot P = P_B \cdot P$, we obtain $P = P \cdot P_{A\cap B}$. On the other hand, $P_C = P_C \cdot P_A = P_C P_B$ implies $P_C = P_C \cdot P$. Therefore, we obtain $P_{A\cap B} = P_{A\cap B} P$ and finally $P_{A\cap B} = P$.]

Remark: In the special case $[P_A, P_B]_- = 0$ of commuting projection operators P_A and P_B, we obtain

$$P_{A\cap B} = P_A \cdot P_B. \tag{1.23}$$

(1.24) THEOREM: The projection operator $P_{A\cup B}$ with the range $M_A \cup M_B$ can be expressed by the projection operators P_A and P_B with the range M_A and M_B, respectively, due to the relation

$$P_{A\cup B} = 1 - \lim_{n\to\infty} \{(1 - P_A)(1 - P_B)\}^n.$$

Proof: It can easily be shown that for subspaces M_A and M_B the relation $M_A \cup M_B = \perp(\perp M_A \cap \perp M_B)$ holds. Therefore, using Theorem (1.22), we get

$$P_{A\cup B} = 1 - P_{\perp A\cap\perp B} = 1 - \lim_{n\to\infty} \{(1 - P_A)(1 - P_B)\}^n.]$$

Remark: In the special case, $[P_A, P_B]_- = 0$ of commuting projection operators P_A and P_B we obtain

$$P_{A\cup B} = P_A + P_B - P_A P_B. \tag{1.25}$$

1.4 STATES AND PROPERTIES OF A PHYSICAL SYSTEM

1.4.1 *The State of a Physical System*

In quantum mechanics, a physical system S – an elementary particle, a nucleus, an atom, etc. – is characterized by its '*state*'. States will be denoted here by lower case Greek letters $\varphi, \psi, \ldots$, and for a system S which is in the state φ we write $S(\varphi)$. In the formal description of a system in quantum theory, a state φ is represented by a vector $\varphi \in \mathcal{H}$ of Hilbert space. The physical meaning of this concept will become clear if, in addition, the notion of an observable is introduced. An observable $\mathscr{A}$, the values of which can be estimated by an experiment, is represented in quantum theory by a self-adjoint operator $A = A[\mathscr{A}]$.[8] The relation between the theory and the experimental observation is then given by the following statement. If the system S is in the state φ, the expectation value $\mathrm{Ex}(\mathscr{A}, S(\varphi))$ of the observable $\mathscr{A}$, which is represented by the operator $A = A[\mathscr{A}]$ is given by

$$\mathrm{Ex}(\mathscr{A}, \varphi) = \langle A \rangle_\varphi = (\varphi, A\varphi). \tag{1.26}$$

From an experimental point of view, the expectation value $\langle A \rangle_\varphi$ is given by the average value of A-measuring results $a_1, a_2, \ldots a_N$, which one obtains if A-measurements are performed on an ensemble $\{S^1(\varphi), S^2(\varphi) \ldots S^N(\varphi)\}$ of N identically prepared systems $S^i(\varphi)$ in the same state φ, where N has to be taken in the limit $N \to \infty$

$$\langle A \rangle_\varphi = \lim_{N\to\infty} \frac{1}{N} \sum_{i=1}^{N} a_i. \tag{1.27}$$

If the operator A has a discrete spectrum, its spectral decomposition has the form $A = \Sigma_i\, a_i P_{a_i}$, where a_i is an eigenvalue and P_{a_i} is the projection operator which projects onto the subspace M_{a_i} which is spanned by the eigenstates $\varphi^\lambda(a_i)$ of the eigenvalue a_i. The possible experimental measuring results are then given by the eigenvalues a_i of the operator A. If, in particular, φ is an eigenstate of A, one obtains the same result $a = a_1 = a_2 \cdots = a_N$ for all systems $S^i(\varphi)$, where a is that eigenvalue of A which satisfies the equation $A\varphi = a\varphi$. This is the case if the projection operator $P[\varphi]$ with range φ commutes with all projection operators P_{a_i}; i.e. if the relations $[P[\varphi], P_{a_i}]_- = 0$ are fulfilled for every i.

According to this interpretation, the state φ of a system S is defined only with respect to an ensemble $\{S^i(\varphi)\}_i$ of equally prepared systems S^i. In principle, the state φ can be obtained experimentally by measuring the expectation values $\langle A^{(i)} \rangle_\varphi$ of a sufficiently large family of observables $\mathscr{A}^{(i)}[A^{(i)}]$. If the expectation values $\langle A \rangle_\varphi$ are known for all self-adjoint operators, the vector $\varphi \in \mathscr{H}$ can be calculated. There is, however, also another possibility to define a state φ with respect to an individual system. For this purpose, we consider a finite or infinite family $\{A^{(i)}\}_i$ of self-adjoint operators which mutually commute, (i.e. $[A^{(i)}, A^{(k)}]_- = 0$), and which commute with $P[\varphi]$, (i.e. $[A^{(i)}, P[\varphi]]_- = 0$). The observables $\mathscr{A}^{(i)}$, which correspond to the operators $A^{(i)}$, can then be measured simultaneously on the system $S(\varphi)$ without influencing its state φ. The measuring results are the eigenvalues $a_\varphi^{(n)}$ of the operators $A^{(n)}$ with respect to the vector φ. This means that for a sufficiently large family $\{A^{(i)}\}_i$ of operators the state φ can be determined by the values $a_\varphi^{(n)}$. Therefore, the state of a system S may be considered as the totality of the measuring values $a_\varphi^{(n)}$ of all operators $A^{(n)}$ which can be measured simultaneously on the system $S(\varphi)$.

In the discussion of the last paragraph, we made use of an important fact concerning the measuring process. If the system S is in the state φ, an operator $A = \Sigma\, a_i P_{a_i}$ can be measured without influencing this state φ if and only if $P[\varphi]$ commutes with all projection operators P_{a_i}. Moreover, two operators A and B can be measured simultaneously if the measuring process of B does not influence the result of a preceding A-measuring result, and *vice versa*. This is the case if and only if the two operators commute; i.e. $[A, B]_- = 0$. With respect to the physical meaning of this property, two operators A and B which satisfy the relation $[A, B]_- = 0$ are also said to be *commensurable*. Thus, in the framework of quantum theory, the concept of *commensurability* is given by a relation between self-adjoint operators. However, this formal definition should be clearly distinguished from the operational interpretation of the concept of commensurability in terms of measuring processes. It is a consequence of the quantum mechanical theory of the measuring process that these two notions of commensurability coalesce.

1.4.2 *Properties of a Physical System*

A projection operator is a bounded linear operator which is self-adjoint ($P = P^*$) and idempotent ($P^2 = P$) (Theorem 1.18). Therefore, a projection operator P may be considered as an operator which corresponds to an observable quantity $\mathscr{P} = \mathscr{P}[P]$. It follows from Theorem (1.18) that a projection operator P_A with range M_A has precisely two eigenvalues 0 and 1, and that these eigenvalues have the following meaning

$$\begin{aligned} P_A f = f &\curvearrowleft\!\curvearrowright f \in M_A \\ P_A f = \mathbf{0} &\curvearrowleft\!\curvearrowright f \in \perp M_A \end{aligned} \tag{1.28}$$

Since the eigenvalues of a self-adjoint operator A correspond to possible measuring values of the observable $\mathscr{A}(A)$, the eigenvalues 0 and 1 (of the operator P_A) correspond to the measuring values of the observable $\mathscr{P}_A = \mathscr{P}_A[P_A]$. Observables which have only two eigenvalues 0 and 1 will be called '*properties*'.

This terminology can be justified by the following arguments. A system $S(\varphi)$ is said to have the *property* P_A if and only if $P_A\varphi = \varphi$; i.e. $\varphi \in M_A$. It is said to have the *property* $P_{\perp A}$ if and only if $P_A\varphi = \mathbf{0}$; i.e. $f \in \perp M_A$. Therefore, the statement '*a system S in the state* φ *has the property* P_A' means that the state-vector φ of the system is an element of that subspace M_A which is the range of the operator $P_A = P_A[\mathcal{P}_A]$. Consequently, properties are represented in the Hilbert space by subspaces. Only in the special case that M_A is a one-dimensional subspace, is the *property* P_A given by a *state* f_A. In this case, the relation $P_A\varphi = \varphi$ means that $\varphi = f_A$. However, in general, '*states*' and '*properties*' are quite different concepts.

It is obvious that the concept of commensurability can also be immediately applied to properties. Two properties $\mathcal{P}_A$ and $\mathcal{P}_B$ are said to be *commensurable* if the two observables can be measured in an arbitrary sequence without thereby influencing the result of the measurement. According to the theory of the measuring process, this is the case if and only if the projection operators P_A and P_B commute; i.e. if $[P_A, P_B]_- = 0$. It is interesting to note that this formal commensurability relation can also be easily expressed in terms of subspaces. This can be done according to the following theorem:

(1.29) THEOREM: Let P_A and P_B be projection operators with range M_A and M_B, respectively. Then the relation $[P_A, P_B]_- = 0$ holds if and only if $M_A = (M_A \cap M_B) \cup (M_A \cap \perp M_B)$.

Proof: If $[P_A, P_B]_- = 0$, it follows by means of (1.23) and (1.25) that $P_A = P\{(A \cap B) \cup (A \cap \perp B)\}$ and therefore $M_A = (M_A \cap M_B) \cup (M_A \cap \perp M_B)$. Since $(M_A \cap M_B)$ and $(M_A \cap \perp M_B)$ are completely orthogonal, an element $f_A = P_A f \in M_A$ can uniquely be decomposed into $f_A = f_A^{(1)} + f_A^{(2)}$ with $f_A^{(1)} \in M_A \cap M_B \subseteq M_B$, $f_A^{(2)} = M_A \cap \perp M_B \subseteq \perp M_B$. Therefore, for any $f \in \mathcal{H}$ the relation $P_A f = P_B P_A f + (1 - P_B) P_A f$ holds. Furthermore, since $P_B P_A f \in M_A$ we have $P_B P_A f = P_A P_B P_A f$, and consequently $[P_A, P_B]_- = 0$.]

In quantum theory, it is sufficient in principle to consider only the most simple type of observables, the '*properties*', since all other observables can be reduced to a set of '*properties*'. In fact, if $A = A^*$ is a self-adjoint operator with a discrete spectrum, the spectral decomposition

$$A = \sum_i a_i P_{a_i} \tag{1.29}$$

expresses the operator A by the projection operators P_{a_i}, which project onto the subspaces M_{a_i}. If $\varphi^\lambda(a_i)$ are the eigenstates of the eigenvalue a_i, the subspace M_{a_i} is spanned by the one-dimensional subspaces $\varphi^\lambda(a_i)$; i.e., $M_{a_i} = \bigcup_\lambda^{N_i} \varphi^\lambda(a_i)$, where N_i is the degeneracy of the eigenvalue a_i. Since the projection operators $P[\varphi^\lambda(a_i)]$ commute and satisfy the relation $P[\varphi^\lambda(a_i)]P[\varphi^\mu(a_i)] = 0$ (for $\lambda \neq \mu$), on account of (1.25) the projection operator P_{a_i} can be expressed by

$$P_{a_i} = \sum_\lambda^{N_i} P[\varphi^\lambda(a_i)].$$

In this way, a self-adjoint operator A can even be reduced to projection operators $P[\varphi]$ which project to one-dimensional subspaces; i.e. to vectors $\varphi \in \mathcal{H}$.

If quantum theory is formulated essentially in terms of properties, it is appropriate to use the following terminology. A physical system S is characterized by its state φ, which is represented by a vector in Hilbert space; i.e. $\varphi \in \mathcal{H}$. The properties $\mathcal{P}_A, \mathcal{P}_B \ldots$ of the system $S(\varphi)$ are then given by all subspaces $M_A, M_B, \ldots$ with $\varphi \in M_A, \varphi \in M_B \ldots$ and by the projection operators $P_A, P_B \ldots$ with $P_A\varphi = \varphi$, $P_B\varphi = \varphi \ldots$, respectively. The statement that 'the system $S(\varphi)$ has the property $\mathcal{P}_A$' will be called the '*proposition*' $A(S, \varphi)$; i.e. we have the following definition:

(1.30) DEFINITION: Proposition $A(S, \varphi) \rightleftharpoons P_A\varphi(S) = \varphi(S) \curvearrowright \varphi(S) \in M_A$.

The concept of proposition will become very important in the following chapters.

Summarizing these considerations, we find that *subspaces* M_A, $M_B \ldots$ correspond to 'properties' $\mathcal{P}_A, \mathcal{P}_B, \ldots$ of a system $S(\varphi)$ and that these properties correspond to 'propositions' $A(S, \varphi), B(S, \varphi), \ldots$. The relation '$\subseteq$' between subspaces will thus lead to a relation between propositions, and the subspaces $M_A \cap M_B$, $M_A \cup M_B$, $\perp M_A$ will correspond to certain compound propositions. The algebra of subspaces, which is given by the lattice $L_H = \langle \mathcal{L}; \subseteq, \perp, \cap, \cup \rangle$, can, in this way, be transferred to propositions. The result is a lattice of quantum mechanical propositions which, from a formal point of view, has many similarities with the propositional lattice of classical logic. The question whether this formal similarity justifies the denotation 'quantum logic' for the lattice will be discussed in the next chapter.

NOTES AND REFERENCES

The theory of Hilbert space is treated in many standard text books. Here, we mention only a few of them which are of particular interest for quantum physics.

[1] P.R. Halmos, *Introduction to Hilbert Space*, Chelsea Publishing Co., New York (1957).

[2] N.J. Achieser and J.M. Glasmann, *Theorie der linearen Operatoren im Hilbert-Raum*, Akademie Verlag, Berlin (1958).

The mathematical foundation of quantum theory can be found in:

[3] J.v. Neumann, *Mathematical Foundations of Quantum Mechanics*, Princeton University Press, Princeton (1955).

[4] G. Ludwig, *Grundlagen der Quantenmechanik*, Springer-Verlag, Berlin (1954).

[5] J.M. Jauch, *Foundations of Quantum Mechanics*, Addison-Wesley Publishing Co., Reading, Mass. (1968).

[6] C. Piron, *Foundations of Quantum Physics*, W.A. Benjamin, Reading, Mass (1976).

The first paper which is concerned with the lattice of subspaces of Hilbert space is:

[7] G. Birkhoff and J.v. Neumann, *Ann. of Math.* **37** (1936) 823.

[8] It is a much debated question whether, conversely, every self-adjoint operator A corresponds to an observable quantity $\mathscr{A} = \mathscr{A}(A)$. We shall assume here that this is actually the case without further discussing the problems of this hypothesis.

CHAPTER 2

THE LOGICAL INTERPRETATION OF THE LATTICE L_q

In this chapter, the lattice of subspaces of a Hilbert space is investigated with respect to its logical interpretation. In Section 2.1, we introduce the abstract lattice L_q, which has as a model the lattice of subspaces of a Hilbert space, and we mention some interesting properties of this lattice. In Section 2.2, the relation of *commensurability* is defined, which is of special interest from a formal point of view as well as for the physical interpretation of L_q. In addition to the operations already defined in L_q, we introduce, in Section 2.3, a further operation, the *material quasi-implication*, the existence of which is of great importance for the logical interpretation of the lattice L_q. Keeping these formal results in mind we shall consider in Section 2.4 the question of what kind of requirements must be fulfilled by a lattice in order that it be interpretable as a logical calculus.

2.1. THE QUASIMODULAR LATTICE L_q

2.1.1. *The Axioms of the Lattice L_q*

For the investigation of the question whether there exists a logical interpretation of the lattice of subspaces of Hilbert space, it is useful to characterize this lattice from an abstract point of view, and to omit all those algebraic structures which have no logical interpretation in the strict sense.[1] The resulting abstract lattice thus obtained will be called L_q. In order to avoid any confusion with the concrete algebra L_H of subspaces, we use a different denotation for the elements as well as for the relation and operations of L_q.

In order to characterize the abstract algebra L_q, we start with a set $\mathcal{S} = \{A, B, C, \dots\}$ of elements $A, B, \dots$ which can be realized by subspaces $M_A, M_B \dots$, respectively. On the set S a 2-place relation $R \subseteq \mathcal{S} \times \mathcal{S}$ is defined, which will be denoted by the sign '$\leq$' and which fulfills the conditions

$L_q(1.1)$ $\quad A \leqslant A,$

$L_q(1.2)$ $\quad A \leqslant B, B \leqslant C \curvearrowright A \leqslant C.$

With respect to this relation, $\mathscr{S}$ is a partially-ordered set. The equivalence relation '=' can then be defined by

$$A = B \quad \text{iff} \quad A \leqslant B \text{ and } B \leqslant A.$$

Furthermore, on the set $\mathscr{S}$ there are two 2-place operations

$$\Theta_\wedge: \mathscr{S} \times \mathscr{S} \to \mathscr{S} \quad \text{denoted by} \quad \Theta_\wedge(A, B) = A \wedge B,$$
$$\Theta_\vee: \mathscr{S} \times \mathscr{S} \to \mathscr{S} \quad \text{denoted by} \quad \Theta_\vee(A, B) = A \vee B,$$

which satisfy the following relations:

$L_q(2.1)$ $\quad A \wedge B \leqslant A$

$L_q(2.2)$ $\quad A \wedge B \leqslant B$

$L_q(2.3)$ $\quad C \leqslant A, C \leqslant B \curvearrowright C \leqslant A \wedge B$

$L_q(3.1)$ $\quad A \leqslant A \vee B$

$L_q(3.2)$ $\quad B \leqslant A \vee B$

$L_q(3.3)$ $\quad A \leqslant C, B \leqslant C \curvearrowright A \vee B \leqslant C$

According to these relations, $A \wedge B$ is the infimum and $A \vee B$ the supremum of A and B with respect to the relation R. Therefore, the algebra $\langle \mathscr{S}; \leqslant, \wedge, \vee \rangle$ is a lattice. In addition, in this lattice there exists a zero element Λ and a unit element V such that for any A

$L_q(4.0)$ $\quad \Lambda \leqslant A, \qquad A \leqslant \mathsf{V}.$

Furthermore, there exists an *automorphism*

$$\Theta_\neg: \mathscr{S} \to \mathscr{S}, \quad \text{denoted by} \quad \Theta_\neg(A) = \neg A,$$

which satisfies the conditions

$L_q(4.1)$ $\quad A \wedge \neg A \leqslant \Lambda,$

$L_q(4.2)$ $\quad \mathsf{V} \leqslant A \vee \neg A,$

$L_q(4.3)$ $\quad A = \neg\neg A,$

$L_q(4.4)$ $\quad A \leqslant B \curvearrowright \neg B \leqslant \neg A,$

so that the lattice considered is *orthocomplemented* and the element $\neg A$ is the *orthocomplement* of A. In addition, this orthocomplemented lattice is *quasimodular*; i.e. the condition

$L_q(5)$ $$B \leqslant A, C \leqslant \neg A \curvearrowright A \wedge (B \vee C) \leqslant B$$

is fulfilled for arbitrary elements A, B and C.

The axioms $L_q(1.1)$–$L_q(5)$ define an algebraic structure which will be denoted here by $L_q = \langle \mathcal{S}; \leqslant, \wedge, \vee, \neg; \Lambda, \mathrm{V}\rangle$ and which is called an *orthocomplemented quasimodular*[2] lattice. The algebra of subspaces of Hilbert space has further and more specific properties than L_q. The lattice L_H is σ-complete, atomic, (1.14), and fulfills the covering law (1.15). These properties are not incorporated into the definition of L_q. If, for example, the lattice L_q is assumed to be σ-complete, this will be mentioned explicitly, and the lattice will then be denoted by $L_q^{(\sigma)}$.

2.1.2. *Some Properties of the Lattice* L_q

In the following, we mention briefly some important properties of the lattice L_q. Thereby we will distinguish between those relations which are already fulfilled in an *orthocomplemented lattice* L_o defined by $L_q(1.1)$–$L_q(4.4)$ and the specific properties of a quasimodular lattice L_q.

(2.1) THEOREM: In an orthocomplemented lattice L_o the following relations (de Morgan's formulae) hold:

$$\text{(a)} \quad \neg(A \wedge B) = \neg A \vee \neg B,$$
$$\text{(b)} \quad \neg(A \vee B) = \neg A \wedge \neg B. \tag{2.2}$$

Remark: The validity of the two de Morgan's formulae shows that in L_o the operation $\Theta_\neg$ is a dual automorphism.

Proof: Using $L_q(4.4)$ we obtain $\neg A \vee \neg B \leqslant \neg(A \wedge B)$ and $\neg(A \vee B) \leqslant \neg A \wedge \neg B$. The inverse relations $\neg(A \wedge B) \leqslant \neg A \vee \neg B$ and $\neg A \wedge \neg B \leqslant \neg(A \vee B)$ follow from $L_q(4.4)$ and $L_q(4.3)$.]

A lattice L with a zero element Λ and a unit element V is called *complemented* if, for any $A \in L$, there exists an element A', the complement of A, which satisfies the conditions

$$A \wedge A' = \Lambda, \qquad A \vee A' = \mathrm{V}.$$

In general, the complement A' is not uniquely defined. This is also the case for an orthocomplemented lattice, since the orthocomplement $\neg A$ is not uniquely defined by the lattice axioms $L_q(4.3)$ and $L_q(4.4)$. Moreover, it is a theorem[3] that if every element in a lattice L has a unique complement A' and if the de Morgan's formulae (2.2) hold, then L is orthocomplemented and *distributive*; (i.e. a *Boolean lattice*). Since the lattice L_q is *not* distributive, it cannot be uniquely complemented. However, in L_q the element $\neg A$ *is in fact* uniquely defined due to the concrete automorphism $\Theta_{\neg}: \mathcal{S} \to \mathcal{S}$, the existence of which has been presupposed with the definition of the lattice L_o.[4] (In the Hilbert space realization, this automorphism is given by the mapping to the completely orthogonal subspace.)

In an orthocomplemented lattice L_o, the quasimodularity $L_q(5)$ is a relaxation of the modularity

$$B \leqslant A \curvearrowright A \wedge (B \vee C) \leqslant B \vee (A \wedge C).$$

On the other hand, the modularity is a weakening of the distributivity. This can easily be seen if we write the three laws in the equivalent formulations

Quasimodularity:

$$B \leqslant A, C \leqslant \neg A \curvearrowright A \wedge (B \vee C) = (A \wedge B) \vee (A \wedge C). \qquad (2.3)$$

Modularity:

$$B \leqslant A \curvearrowright A \wedge (B \vee C) = (A \wedge B) \vee (A \wedge C). \qquad (2.4)$$

Distributivity:

$$A \wedge (B \vee C) = (A \wedge B) \vee (A \wedge C). \qquad (2.5)$$

It is obvious that, according to these reformations, counter-examples can be constructed which show that not every quasimodular lattice is also modular, and that not every modular lattice is also distributive.

In addition, it should be mentioned that the quasimodular law is equivalent to its dual formulation.

(2.6) THEOREM: In an orthocomplemented lattice L_o, the two laws:

$$B \leqslant A, C \leqslant \neg A \curvearrowright A \wedge (B \vee C) = B, \qquad (2.7)$$

$$A \leqslant B, \neg A \leqslant C \curvearrowright A \vee (B \wedge C) = B, \qquad (2.8)$$

are equivalent.

Proof: If (2.7) is assumed to be generally valid, from $A \leqslant B$, $\neg A \leqslant C$ it follows by means of $L_q(4.4)$ that $\neg B \leqslant \neg A$, $\neg C \leqslant A$ holds, and therefore $\neg A \wedge (\neg B \vee \neg C) = \neg B$. Using (2.2) one finally obtains $A \vee (B \wedge C) = B$. The inverse proof is analogous.

Instead of the *quasimodular law* (2.7) and (2.8), respectively, one often uses *weak modularity*, which can be shown to be equivalent to it.

(2.9) THEOREM: In an orthocomplemented lattice L_o, the quasimodular law (2.7) is necessary and sufficient for *weak modularity*

$$B \leqslant A \curvearrowright A \wedge (B \vee \neg A) = B \tag{2.10}$$

to hold.

Proof: (a) If we put $C = \neg A$ in (2.7), one obtains (2.10). (b) If we assume $B \leqslant A$, $C \leqslant \neg A$, then we get $B \vee C \leqslant B \vee \neg A$ and $A \wedge (B \vee C) \leqslant A \wedge (B \vee \neg A)$. Using (2.10), it follows that $A \wedge (B \vee C) \leqslant B$. Furthermore, from $B \leqslant A$ one obtains $B \leqslant A \wedge (B \vee C)$, completing the proof.]

According to Theorem (2.6) the *quasimodular law* is equivalent to its dual form. A similar result can be proved for *weak modularity*.

(2.11) THEOREM: In an orthocomplemented lattice L_o weak modularity (2.10) and its dual form

$$A \leqslant B \curvearrowright A \vee (B \wedge \neg A) = B \tag{2.12}$$

are equivalent.

Proof: From $A \leqslant B$ it follows that $\neg B \leqslant \neg A$ and by means of (2.10) one obtains $\neg A \wedge (\neg B \vee \neg\neg A) = \neg B$. Using the de Morgan's formulae (2.2) one finally obtains $A \vee (B \wedge \neg A) = B$; i.e., (2.12). By an analogous argument, one shows that (2.12) implies (2.10).]

Summarizing these results, we find that, in an orthocomplemented lattice L_o, the four laws

(1)	$B \leqslant A, C \leqslant \neg A \curvearrowright A \wedge (B \vee C) = B$	Quasimodularity
(2)	$A \leqslant B, \neg A \leqslant C \curvearrowright A \vee (B \wedge C) = B$	dual form of (1)
(3)	$B \leqslant A \curvearrowright A \wedge (B \vee \neg A) = B$	Weak Modularity
(4)	$A \leqslant B \curvearrowright A \vee (B \wedge \neg A) = B$	dual form of (3)

are equivalent.

2.2 THE RELATION OF COMMENSURABILITY

For the physical interpretation of the lattice L_q and for a more detailed investigation of its formal structure, it is interesting to introduce here the concept of commensurability. For this purpose, we start again from an orthocomplemented lattice L_o and define in this lattice a 2-place relation $K \leqslant L_o \times L_o$, which will be called the *commensurability relation*:

(2.13) DEFINITION: In an orthocomplemented lattice L_o, the 2-place relation $K \subseteq L_o \times L_o$ (commensurability) is defined by

$$(A, B) \in K \curvearrowleft\!\!\curvearrowright A = (A \wedge B) \vee (A \wedge \neg B),$$

and will be denoted by $A \sim B$.

The name 'commensurability' for the relation K can be motivated by considering the realization of the lattice L_o by subspaces of a Hilbert space. In fact, it follows from Theorem (1.29) that the relation K holds for subspaces M_A and M_B if and only if the corresponding projection operators P_A and P_B commute; i.e., if the observables $\mathcal{P}_A$ and $\mathcal{P}_B$ are commensurable in the usual sense of quantum physics. Here, however, we are not concerned with subspaces, and the relation K must be considered as a purely abstract relation defined by (2.13).

It follows from the definition (2.13) that the partial-ordering relation R is contained in the relation K; i.e.

$$R \subseteq K \subseteq L_o \times L_o. \tag{2.14}$$

In fact, it follows from $A \leqslant B$ by means of $L_q(2.3)$ that $A \leqslant A \wedge B$, which implies $A \leqslant (A \wedge B) \vee (A \wedge \neg B)$. The inverse, (i.e. $(A \wedge B) \vee (A \wedge \neg B) \leqslant A$) is always true in L_o.

In an orthocomplemented lattice L_o the relation K is, in general, not symmetric. The symmetry of the commensurability relation is rather a condition which in L_o is equivalent to the quasimodularity. In fact we have the following result[9,10]:

(2.15) THEOREM: In an orthocomplemented lattice L_o, the relation K is symmetric; i.e. $A \sim B \curvearrowright B \sim A$ if and only if the lattice is quasimodular, or equivalently, $L_q(5)$ holds.

Remark: According to its physical meaning the relation of commensurability should be symmetric. Hence, it is obvious that in the lattice L_H the relation K is in fact symmetric. However, the interesting content of this theorem is that it is precisely the quasimodular law which is necessary and sufficient for the symmetry of K. Thus, the theorem illustrates the importance of the Axiom $L_q(5)$. For the proof we use the equivalent form (2.12).

Proof: (a) (Quasimodularity implies symmetry.) If one assumes $A \sim B$; i.e.

$$A = (A \wedge B) \vee (A \wedge \neg B),$$

it follows that

$$\begin{aligned} B \wedge \neg A &= B \wedge \neg(A \wedge B) \wedge (\neg A \vee B) \\ &= B \wedge \neg(A \wedge B). \end{aligned}$$

Applying (2.12), one obtains

$$B = (A \wedge B) \vee (\neg A \wedge B) \quad \text{i.e. } B \sim A.]\!]$$

(b) (Symmetry implies quasimodularity.) In order to prove (2.12), we assume $A \leqslant B$. Then it follows that $A \sim B$, and using the symmetry of K one obtains $B \sim A$; i.e.

$$B = (B \wedge A) \vee (B \wedge \neg A) = A \vee (B \wedge \neg A).]\!]$$

In order to further explain the meaning of the commensurability relation, it is interesting to investigate the relation between commensurability and distributivity. The lattice L_q is not distributive. I.e., the distributive law $A \wedge (B \vee C) = (A \wedge B) \vee (A \wedge C)$ does not hold generally in the quasimodular lattice L_q. However, if the elements B and C are both commensurable with A, distributivity can be demonstrated.[11,12]

(2.16) THEOREM: In an orthocomplemented quasimodular lattice L_q the law of 'weak distributivity', i.e.

$$B \sim A, C \sim A \curvearrowright A \wedge (B \vee C) = (A \wedge B) \vee (A \wedge C),$$

is valid.

Proof: From the premises

$$B = (B \wedge A) \vee (B \wedge \neg A), \qquad C = (C \wedge A) \vee (C \wedge \neg A)$$

it follows that

$$A \wedge (B \vee C) = A \wedge \{(B \wedge A) \vee (B \wedge \neg A) \vee (C \wedge A) \vee (C \wedge \neg A)\}.$$

Applying the quasimodular law to the right-hand side, we obtain

$$A \wedge (B \vee C) = (A \wedge B) \vee (A \wedge C).⟧$$

The meaning of *weak distributivity* is further illustrated by the fact that the relation of commensurability is closed with respect to the lattice operations $\wedge$, $\vee$ and $\neg$.

(2.17) THEOREM: In the lattice L_q, the relation K is closed under the operations $\wedge$, $\vee$, $\neg$; i.e.

(a) $A \sim B, A \sim C \curvearrowright A \sim (B \wedge C)$

(b) $A \sim B, A \sim C \curvearrowright A \sim (B \vee C)$

(c) $A \sim B \curvearrowright A \sim \neg B$

Proof: $A \sim B$ means $A = (A \wedge B) \vee (A \wedge \neg B)$. Thus (c) is obvious. Assuming $A \sim B$ and $A \sim C$, it follows, due to the symmetry of K, that

$$B \vee C = \{(B \wedge A) \vee (C \wedge A)\} \vee \{(B \wedge \neg A) \vee (C \wedge \neg A)\}.$$

Applying weak distributivity (2.16) to the two brackets of the right-hand side, one obtains

$$B \vee C = \{A \wedge (B \vee C)\} \vee \{\neg A \wedge (B \vee C)\}.$$

I.e. $B \vee C \sim A$, and thus $A \sim (B \vee C)$, proving (b). To prove (a), from the assumption $A \sim B$, $A \sim C$, we first conclude $A \sim \neg B$, $A \sim \neg C$ (by means of (c)). Applying (b) we obtain $A \sim \neg B \vee \neg C$ and again using (c) $A \sim \neg(\neg B \vee \neg C) = B \wedge C$, completing the proof of (a).⟧

It follows from this closure property of the relation K, together with the above mentioned 'weak distributivity' (Theorem 2.16), that three elements $A, B, C \in L_q$ which are pairwise commensurable will generate a sublattice $L(A, B, C) \subseteq L_q$, which is *orthocomplemented* and *distributive* (i.e. a *Boolean* sublattice $L_B(A, B, C) \subseteq L_q$). Conversely – since in a Boolean lattice the relation $A \sim B$ holds for all elements – in a Boolean sublattice $L_B \subseteq L_q$, any two elements $A, B \in L_B$ will be commensurable; i.e. $A \sim B$ will be valid.

Summarizing these results concerning commensurability and Boolean sublattices, we arrive at the following statement: In the lattice L_q, the relation $K \subseteq L_q \times L_q$ fulfills four conditions:

(K1) K is symmetric.

(K2) $R \subseteq K$.

(K3) If $\mathcal{S} \subseteq L_q$ is a subset of elements with $\mathcal{S} \times \mathcal{S} \subseteq K$, then $\mathcal{S}$ generates a Boolean sublattice $L_B(\mathcal{S}) \subseteq L_q$.

(K4) If $L_B \subset L_q$ is a Boolean sublattice, the elements of any subset $\mathcal{S} \subseteq L_B$ are commensurable; i.e., $\mathcal{S} \times \mathcal{S} \subseteq K$.

Conversely, it can be shown that these conditions are also sufficient in order to characterize K as an abstract relation.[13]

(2.18) THEOREM: In a lattice L_q the relation $K \subseteq L_q \times L_q$ is *uniquely* defined by the conditions (K1)–(K4).

Proof: Let K and K' be relations which both satisfy (K1)–(K4). If $\mathcal{S} \subseteq L_q$ and $\mathcal{S} \times \mathcal{S} \subseteq K$ then $\mathcal{S}$ generates a Boolean sublattice $L_B(\mathcal{S})$ (on account of (K3)). Since $L_B(\mathcal{S})$ is a Boolean sublattice, it follows (due to (K4)) that $L_B(\mathcal{S}) \times L_B(\mathcal{S}) \subseteq K'$. Hence, for any $\mathcal{S} \subseteq L_q$, the relation $\mathcal{S} \times \mathcal{S} \subseteq K$ implies $\mathcal{S} \times \mathcal{S} \subseteq L_B(\mathcal{S}) \times L_B(\mathcal{S}) \subseteq K'$. Therefore, one obtains for all $\mathcal{S}^{(n)} \subseteq L_q$: $\bigcup_n \mathcal{S}^{(n)} \times \mathcal{S}^{(n)} \subseteq K'$. On the other hand, it follows from (K1) and (K2) that for any pair $A, B \in L_q$: $(A, B) \in K$ implies $(A, B) \times (A, B) \subseteq K$. Therefore we obtain $K \subseteq \bigcup_n \mathcal{S}^{(n)} \times \mathcal{S}^{(n)}$ and thus $K \subseteq K'$. By an analogous argument it follows $K' \subseteq K$.]

COROLLARY: If $L_B^{(n)}$ $(n = 1, 2, \ldots)$ are the Boolean sublattices of L_q, it follows from Theorem (2.18) that the relation K can be expressed by $K = \bigcup_n (L_B^{(n)} \times L_B^{(n)})$.

According to the results thus obtained, in a lattice L_q there exists one and only one relation K which satisfies (K1)–(K4). In addition, the inverse statement can also be proved.

(2.19) THEOREM: An orthocomplemented lattice L_o, in which there exists a relation K satisfying (K1)–(K4), is quasimodular.

Proof: Let $A, B, C \in L_o$ be three elements such that $B \leq A$, $C \leq \neg A$. Thus the premises of (2.7) are fulfilled. Then it follows due to (K1) and (K2) that $A \sim B$, $A \sim \neg C$, $B \sim \neg C$. Hence, the lattice $L(A, B, \neg C)$ generated by $A, B, \neg C$ is Boolean and in particular one obtains $A \wedge (B \vee C) = (A \wedge B) \vee (A \wedge C) = B$.]

The connection between the relation of commensurability and quasimodularity is thus fully established. If, in a lattice L_o, there exists a relation K satisfying (K1)–(K4), then this lattice is quasimodular. Conversely, if a lattice L_o is quasimodular, then there exists a relation K – given by $A \sim B$ – which satisfies (K1)–(K4).

In an orthocomplemented quasimodular lattice L_q, the relation

$K \subseteq L_q \times L_q$ can also be expressed by a 2-place operation Θ_k: $L_q \times L_q \to L_q$ denoted by $\Theta_k(A, B)$, in the following way:

(2.20) THEOREM: In a lattice L_q, for any two elements $A, B \in L_q$, there exists an element $k(A, B)$ which satisfies the condition

$$A \sim B \leftrightarrow V \leq k(A, B) \tag{2.21}$$

and which is given by $k(A, B) = k_0(A, B)$, where

$$k_0(A, B) = (A \wedge B) \vee (A \wedge \neg B) \vee (\neg A \wedge B) \vee (\neg A \wedge \neg B). \tag{2.22}$$

Proof: (a) If $A \sim B$ we have $A = (A \wedge B) \vee (A \wedge \neg B)$, and by means of the symmetry of $K, \neg A = (\neg A \wedge B) \vee (\neg A \wedge \neg B)$. Hence, we obtain

$$V = A \vee \neg A = (A \wedge B) \vee (A \wedge \neg B) \vee (\neg A \wedge B) \vee (\neg A \wedge \neg B).$$

(b) If $V = (A \wedge B) \vee (A \wedge \neg B) \vee (\neg A \wedge B) \vee (\neg A \wedge \neg B)$, one obtains

$$A = A \wedge \{[(A \wedge B) \vee (A \wedge \neg B)] \vee [(\neg A \wedge B) \vee (\neg A \wedge \neg B)]\}.$$

Applying the quasimodular law (2.7) to the right-hand side it follows that

$$A = (A \wedge B) \vee (A \wedge \neg B) \quad \text{(i.e. } A \sim B).]\!]$$

Remark: This proof shows that the implication $A \sim B \rightarrow V \leq k(A, B)$ is valid even in an orthocomplemented lattice L_o. However, for the proof of $V \leq k(A, B) \rightarrow A \sim B$ the quasimodular law is indispensable.

By means of the explicit form

$$k(A, B) = (A \wedge B) \vee (A \wedge \neg B) \vee (\neg A \wedge B) \vee (\neg A \wedge \neg B)$$

of the commensurability operation $k(A, B)$, some properties of $k(A, B)$ can be derived, which will prove to be of particular interest for the logical interpretation of the lattice L_q.

(2.23) THEOREM: In a lattice L_q the element $k(A, B) = k_o(A, B)$ has the following properties:

(a) $k(A, B) = k(\neg A, B) = k(B, A)$

(b) $A \wedge B \leq k(A, B)$

(c) $A \wedge k(A, B) \leq (A \wedge B) \vee (A \wedge \neg B)$

(d) $k(A, B) = (k(A, B) \wedge A) \vee (k(A, B) \wedge \neg A)$

Proof: (a) and (b) are obvious. (c) can be proved by applying the quasimodular law (2.7) to $A \wedge k(A, B)$. For the proof of (d) one must apply (2.7) to $k(A, B) \wedge A$ and to $k(A, B) \wedge \neg A$.

2.3 THE MATERIAL QUASI-IMPLICATION

A Boolean lattice L_B is defined as an orthocomplemented and distributive lattice. Hence for an axiomatisation of L_B one has to add to the axioms $L_q(1.1)$–$L_q(4.4)$ a further axiom. Whereas we added $L_q(5)$ in order to obtain quasimodular lattices, we now add $L_B(5)$ in order to obtain Boolean lattices: To wit

$$L_B(5) \qquad A \wedge (B \vee C) = (A \wedge B) \vee (A \wedge C).$$

In a Boolean lattice the partial-ordering relation R can also be defined by a 2-place operation $\Theta_{\rightarrow}$: $L_B \times L_B \rightarrow L_B$ in the following way: For any two elements $A, B \in L_B$ there exists an element $A \rightarrow B$ which satisfies the conditions

$$A \wedge (A \rightarrow B) \leqslant B \tag{2.24}$$

$$A \wedge C \leqslant B \curvearrowright C \leqslant A \rightarrow B. \tag{2.25}$$

The element $A \rightarrow B$ is *uniquely* determined by these conditions and is given by

$$A \rightarrow B = \neg A \vee B. \tag{2.26}$$

It then follows from the axioms of the lattice L_B that this element is connected with the partial-ordering relation R by the condition

$$A \leqslant B \curvearrowright V \leqslant A \rightarrow B. \tag{2.27}$$

In the logical interpretation of the lattice L_B the *relation* '$\leqslant$' is called 'implication' and correspondingly the *operation* $A \rightarrow B$ 'material implication'. In this interpretation – which will be discussed in more detail in the next section – condition (2.27) states that $A \rightarrow B$ is 'true' if and only if $A \leqslant B$ 'holds' and condition (2.24) corresponds to the 'modus ponens' law, which is essential for any logical inference.[14]

The existence of a *material implication* $A \rightarrow B$ is very closely connected with the distributive law. In fact, the following theorem holds: If in an orthocomplemented lattice L_o for any two elements $A, B \in L_o$ there exists an element $A \rightarrow B$ which satisfies (2.24) and (2.25), then the distributive law $L_B(5)$ is valid and, hence, the lattice is

Boolean. Therefore, we find that the existence of an element $A \to B$, which satisfies (2.24) and (2.25) and which is very important for the logical interpretation, is *necessary* and *sufficient* for the distributivity of the lattice L_o considered here.

We have mentioned here these well-known properties of a Boolean lattice, since in the lattice L_q we are confronted with a very similar situation. For a logical interpretation one must again require the existence of an element $A \to B$ which satisfies at least the conditions (2.24) and (2.26). It is obvious that the condition (2.25), which is of minor importance for the logical interpretation, cannot be fulfilled since in that case the lattice would be distributive. However, it turns out that a somewhat weaker condition can be formulated which fulfills all the necessary requirements.[15]

In fact, also in a lattice L_q, the partial-ordering relation R can be defined by a 2-place operation $\Theta_{\to}$: $L_q \times L_q \to L_q$ in the following way:

(2.28) THEOREM: In an orthocomplemented quasimodular lattice L_q, for any two elements $A, B \in L_q$ there exists an element $A \to B$ which satisfies the conditions

$$A \wedge (A \to B) \leqslant B, \tag{2.29}$$

$$A \wedge C \leqslant B \curvearrowright \neg A \vee (A \wedge C) \leqslant A \to B, \tag{2.30}$$

and which is given by

$$A \to B = \neg A \vee (A \wedge B). \tag{2.31}$$

Proof: (2.29): $A \wedge (A \to B) = A \wedge (\neg A \vee (A \wedge B)) = A \wedge B \leqslant B$, using the quasimodular law.

(2.30): From $A \wedge C \leqslant B$ it follows that $A \wedge C \leqslant A \wedge B$ and hence

$$\neg A \vee (A \wedge C) \leqslant \neg A \vee (A \wedge B) = A \to B.]\!]$$

It is obvious that the conditions (2.29), (2.30) and (2.31) are relaxations of the conditions (2.24), (2.25) and (2.26), respectively, which are satisfied in a Boolean lattice L_B. In fact, in an orthocomplemented lattice L_o the conditions (2.24) and (2.25) imply the weaker conditions (2.29) and (2.30). Since, in addition, distributivity follows from (2.24) and (2.25), we have $\neg A \vee B = \neg A \vee (A \wedge B)$ and thus the two elements (2.26) and (2.31) agree. On the other hand, in L_o, the conditions (2.30), (2.31) do not imply (2.25) and (2.26), since otherwise the lattice L_q would be distributive.

Furthermore, it follows from the conditions (2.29) and (2.30) that the element $A \to B = \neg A \vee (A \to B)$ satisfies in L_q the condition

$$A \leqslant B \Leftrightarrow V \leqslant A \to B, \tag{2.32}$$

which is identical with (2.27). Thus, we find that also in the quasi-modular lattice L_q the partial-ordering relation R can be defined by an operation $A \to B$, which is given here by $\neg A \vee (A \to B)$. Since the element $A \to B = \neg A \vee (A \wedge B)$ fulfills conditions (2.29) and (2.32) and at least the weaker condition (2.30) (compared with (2.26)), it has almost the properties of the material implication in L_B, and will therefore be called '*material quasi-implication*'.

In addition to these statements, it can be shown that in the lattice L_q the *material quasi-implication* is even *uniquely* determined by the conditions (2.29) and (2.30) just as the *material implication* is uniquely defined in a Boolean lattice by (2.24) and (2.25).

(2.33) THEOREM: In a lattice L_q there exists *only one* element $A \to B$ which satisfies the relations (2.29) and (2.30).

Proof: According to Theorem (2.28) there exists one element $A \to B = \neg A \vee (A \wedge B)$ which satisfies (2.29) and (2.30). In order to prove the uniqueness, let us assume that there is still another element $q(A, B)$ which also satisfies (2.29) and (2.30). If we put $C = B$ we obtain from (2.30) $\neg A \vee (A \wedge B) \leqslant q(A, B)$ and thus

$$A \to B \leqslant q(A, B). \tag{2.34}$$

Furthermore from (2.29) we obtain

$$A \wedge q(A, B) \leqslant A \wedge B \leqslant \neg A \vee (A \wedge B)$$

and hence

$$\neg A \vee (q(A, B) \wedge A) \leqslant A \to B. \tag{2.35}$$

From (2.34), we obtain $\neg A \leqslant q(A, B)$. Therefore, we can apply the quasimodular law (2.8) to (2.35) and thus obtain

$$q(A, B) \leqslant A \to B$$

completing the proof.]

In an orthocomplemented quasimodular lattice the conditions (2.29) and (2.30) are not only fulfilled by the material quasi-implication, they are even the strongest conditions which can be postulated for $A \to B$.

As was the case in a Boolean lattice, the existence of the material quasi-implication implies the quasimodularity of the lattice.

(2.36) THEOREM: An orthocomplemented lattice L_o, with the property that for any two elements $A, B \in L_o$ there exists an element $A \to B$ which satisfies the conditions (2.29) and (2.30), is quasimodular, i.e. the quasimodular law (2.7)

$$B \leq A, C \leq \neg A \curvearrowright A \wedge (B \vee C) \leq B$$

holds.

Proof: From the premises we have $A \wedge B \leq B$ and $A \wedge C \leq B$ and with (2.30)

$$\neg A \vee (A \wedge B) \leq A \to B,$$
$$\neg A \vee (A \wedge C) \leq A \to B.$$

On account of

$$B \leq B \wedge A \leq \neg A \vee (A \wedge B) \leq A \to B,$$
$$C \leq \neg A \leq \neg A \vee (A \wedge B) \leq A \to B,$$

it follows that $B \vee C \leq A \to B$ and with (2.29) finally

$$A \wedge (B \vee C) \leq A \wedge (A \to B) \leq B.$$]

This theorem gives rise to an alternative axiomatisation of the lattice L_q. Instead of the quasimodular law $L_q(5)$ one could require for all pairs $A, B \in L_o$ the existence of an element $A \to B$ which satisfies the axioms

$L_q(5.1)$ $\quad A \wedge (A \to B) \leq B,$

$L_q(5.2)$ $\quad A \wedge C \leq B \curvearrowright \neg A \vee (A \wedge C) \leq A \to B.$

This axiomatisation is of particular interest for the logical interpretation of the lattice L_q.

Concluding this section, we mention here briefly some properties of the material quasi-implication which will become interesting for the logical interpretation of L_q.

(2.37) THEOREM: In the lattice L_q the quasi-implication has the properties:

(a) $\quad A \vee (A \to B) = \mathrm{V}$ $\quad$ (2.38)

(b) $\quad (A \to B) \to A \leq A$ $\quad$ (Peirce's law) $\quad$ (2.39)

Remark: These two relations are known to be valid in a Boolean lattice for the material implication $\neg A \vee B$.[16]

Proof:

(a) $A \vee (A \to B) = A \vee (\neg A \vee (A \wedge B)) = (A \vee \neg A) \vee (A \wedge B) = \mathsf{V}$

(b) $(A \to B) \to A = \neg(A \to B) \vee (A \wedge (A \to B))$
$= (A \wedge \neg(A \wedge B)) \vee (A \wedge (A \to B) \leqslant A.$]

The connection between the material quasi-implication $A \to B$ and the commensurability relation K can be established by many relations. Here we mention only two of them.

(2.40) THEOREM: In the lattice L_q the following relations hold:

(a) $A \sim B$ implies $A \leqslant B \to A$, (2.41)

(b) $A \sim (A \to B)$. (2.42)

Proof:

(a) $A = (A \wedge B) \vee (A \wedge \neg B) \leqslant \neg B \vee (A \wedge B) = B \to A$

(b) $A \to B = \neg A \vee (A \wedge B) = (\neg A \wedge (A \to B)) \vee (A \wedge B \wedge (A \to B))$
$\leqslant (A \wedge (A \to B)) \vee (\neg A \wedge (A \to B).$

Using the symmetry of K we finally obtain $A \sim (A \to B)$.]

In a Boolean lattice, the three 2-place operations $\wedge$, $\vee$ and $\to$ can in each case be expressed by one of them and the operation $\neg$; i.e. by $(\wedge, \neg)$, $(\vee, \neg)$ and $(\to, \neg)$, respectively. This follows from the de Morgan formulae (2.2), the explicit expression (2.26) of the material implication and the inversion $A \vee B = \neg B \to A$ of this formula.

In an orthocomplemented quasimodular lattice L_q, we find a similar situation. The lattice operations $\wedge$, $\vee$ and $\to$ can again be expressed by the operations $(\wedge, \neg)$, $(\vee, \neg)$ and $(\to, \neg)$, respectively.

(2.43) THEOREM: In the lattice L_q the operations $\wedge$, $\vee$, and $\to$ can be expressed in terms of $(\wedge, \neg)$, $(\vee, \neg)$ and $(\to, \neg)$, respectively, according to

$(\wedge, \neg)$		$(\vee, \neg)$		$(\to, \neg)$
$A \wedge B$	$=$	$\neg(\neg A \vee \neg B)$	$=$	$\neg((A \to B) \to \neg A)$
$\neg(\neg A \wedge \neg B)$	$=$	$A \vee B$	$=$	$(\neg A \to \neg B) \to A$
$\neg(A \wedge \neg(A \wedge B))$	$=$	$\neg A \vee \neg(\neg A \vee \neg B)$	$=$	$A \to B$

Proof: It is sufficient to show that $A \vee B = (\neg A \to \neg B) \to A$. The other relations then follow by means of (2.31) and the de Morgan formulae (2.2). According to the weak modularity (2.12), we have

$$A \vee B = A \vee [\neg A \wedge (A \vee B)]$$

and thus

$$\begin{aligned} A \vee B &= [A \wedge (A \vee (\neg A \wedge \neg B))] \vee [\neg A \wedge \neg(\neg A \wedge \neg B)] \\ &= \neg[A \vee (\neg A \wedge \neg B)] \vee [A \wedge (A \vee (\neg A \wedge \neg B))] \\ &= [A \vee (\neg A \wedge \neg B)] \to A = (\neg A \to \neg B) \to A. \rrbracket \end{aligned}$$

2.4 THE RELATION BETWEEN LATTICE THEORY AND LOGIC

In the framework of quantum theory there exists a model for the lattice L_q. The elements $A, B, \ldots \in L_q$ can be interpreted as propositions $A(S, \varphi) \rightleftharpoons P_A\varphi(S) = \varphi(S)$ about a quantum mechanical system S which is in the state $\varphi(S)$. Therefore, the lattice L_q may be considered as a propositional lattice. The partial-ordering relation '$\leq$' will thus lead to a relation between propositions, and the operations $A \wedge B$, $A \vee B$, $A \to B$ and $\neg A$ will correspond to certain compound propositions, the conceptual meaning of which is not yet clear. Furthermore, there are some formal similarities between the lattice L_q and lattices which have logical calculi as models. As examples, we mention here the Boolean lattice L_B of classical logic and the implicative lattice L_i of intuitionistic logic. Here, the lattice operations $\wedge$, $\vee$, $\to$ and $\neg$ correspond to the logical connectives 'and', 'or', 'then' and 'not', respectively. Therefore, it could be conjectured that a logical calculus can also be found which is a model for L_q and which, for that reason, could be called 'quantum logic'.

In order to treat the question whether there is a logical interpretation of the lattice L_q in an adequate manner, we first consider the more general problem of what kind of properties must be required by a lattice such that it can be interpreted as a logical calculus. It will turn out that there are two kinds of fundamental postulates which must be fulfilled by any lattice interpretable as a logical calculus and which we will call here *syntactic* and *semantic requirements*. These postulates will be illustrated by the two examples mentioned, the Boolean lattice L_B and the implicative lattice L_i.[17]

A lattice L_i is called *implicative* – or *relatively pseudo-*

complemented, – if, for any two elements $A, B \in L_i$, there exists an element $A \to B$, the *material implication*, which satisfies the conditions

$$A \wedge (A \to B) \leqslant B, \tag{2.44}$$

$$A \wedge C \leqslant B \curvearrowright C \leqslant A \to B. \tag{2.45}$$

If there is a zero element Λ in L_i, an element $\neg A = A \to \Lambda$ – the *pseudocomplement* – can be defined which satisfies the relation $A \wedge \neg A \leqslant \Lambda$ but, in general, not the relation $V \leqslant A \vee \neg A$. An important property of an implicative lattice is that it is *distributive*. Furthermore, the element $A \to B$ is uniquely defined by the Axioms (2.44), (2.45), just as in a Boolean lattice, but in L_i the material implication cannot be expressed by the other operations $\wedge$, $\vee$ and $\neg$. It is obvious that an implicative lattice L_i is a generalization of a Boolean lattice which no longer contains the relation $V \leqslant A \vee \neg A$ but which is still distributive. In the logical interpretation, the relation $V \leqslant A \vee \neg A$ is called *tertium non datur.*

2.4.1 *Syntactic Requirements*

In a lattice L with a logical interpretation for any element $A \in L$ there must be an element $\neg A \in L$, which can be considered as the *negation* of A. In order to express merely the fact that a proposition $A \in L$ is false, it would be sufficient to require the existence of a zero element Λ and to write $A \leqslant \Lambda$. However, in a logical calculus it should be possible to iterate the negation and to combine it with the other logical connectives, for instance, in the *law of contradiction* $A \wedge \neg A \leqslant \Lambda$. Therefore one must require the existence of a lattice element $\neg A$ which satisfies the law of contradiction and which is related to the Λ-element by

$$V \leqslant \neg A \curvearrowleft\!\curvearrowright A \leqslant \Lambda. \tag{2.46}$$

In a Boolean lattice it is obvious that these two postulates are fulfilled by the *orthocomplement* $\neg A$. In an implicative lattice L_i with zero element Λ the negation can be expressed by the *pseudocomplement* $\neg A = A \to \Lambda$. In fact, the law of contradiction follows from this definition and relation (2.44) if one puts $B = \Lambda$. Moreover, it follows from (2.44) and (2.45) for $B = \Lambda$ that the pseudocomplement is uniquely defined by these relations and fulfills the condition (2.46).

Furthermore, in a lattice L with a logical interpretation it must be possible to define the partial-ordering relation '$A \leq B$' by a 2-place operation '$A \to B$'. The relation $A \leq B$ is interpreted in a logical calculus as the *implication* 'A implies B'. Since, in logical systems, the implication $A \leq B$ is often used in iterated form, for instance in the *modus ponens law*, it must be required that for any two elements $A, B \in L$ there exists an element $A \to B$, such that the modus ponens law can be expressed in the form

$$A \wedge (A \to B) \leq B \qquad \text{(modus ponens law)} \tag{2.47}$$

and that it be connected to the relation $A \leq B$ by the condition

$$\mathrm{V} \leq A \to B \Leftrightarrow A \leq B. \tag{2.48}$$

These requirements are obviously fulfilled in a Boolean lattice L_B, where the *material implication* $A \to B = \neg A \vee B$ is uniquely determined by (2.24), (2.25) and satisfies the conditions (2.47), (2.48). They are also fulfilled in an implicative lattice L_i. Here, the element $A \to B$ – the *relative pseudocomplement* of A in B – is again uniquely defined by (2.44), (2.45) and satisfies the conditions (2.47), (2.48).

The two formal (syntactic) postulates concerning the operations $\neg A$ and $A \to B$ which we have formulated here are easily seen to be fulfilled also by the quasimodular lattice L_q. In fact, in L_q the orthocomplement $\neg A$, which is determined by the axioms $L_q(4.1)$–$L_q(4.4)$, satisfies the relation $A \wedge \neg A \leq \Lambda$ (law of contradiction) and the condition (2.46). Furthermore, the material quasi-implication $A \to B$ is given by $\neg A \vee (A \wedge B)$ and satisfies – according to (2.29) and (2.32) – the conditions (2.47) (modus ponens law) and (2.48), respectively. Therefore, from a purely syntactical point of view there are no evident objections against a logical interpretation of the quasimodular lattice L_q.

2.4.2 *Semantic Requirements*

The syntactic requirements just mentioned must be considered as *necessary* minimal postulates which have to be fulfilled by any lattice with a logical interpretation, but which are, however, by no means *sufficient*. More important – and also in some sense sufficient for the logical interpretation – are the semantic requirements which must be fulfilled by a lattice. By that we mean the existence of a logical semantic, which can be applied to the lattice considered. Here, we consider two different semantics, first the truth-value semantics and second the operational semantics which makes use of dialogs.

A Boolean lattice L_B can be interpreted by a two-valued truth function. To every element $A \in L_B$ one of the two possible truth values 1 (true) and 0 (false) can be assigned; i.e. there exists a *truth-function*

$$t: L_B \to \{0, 1\} \tag{2.49}$$

which satisfies the following conditions:

$$(T_1) \quad t(A) \in \{0, 1\},\ t(\Lambda) = 0,\ t(V) = 1 \tag{2.50}$$

$$(T_2) \quad A \leq \neg B \curvearrowright t(A \vee B) = t(A) + t(B) \tag{2.51}$$

$$(T_3) \quad t(A) = 1,\ t(B) = 1 \curvearrowright t(A \wedge B) = 1. \tag{2.52}$$

The meaning of this truth-function consists in the fact that every element A of a propositional lattice L_B (i.e. every proposition) can be considered as either true (value 1) or false (value 0). Propositions which have this property are said to be *value-definite.* The three conditions (T_1), (T_2), (T_3) further specify the meaning of the concept of truth. For a Boolean lattice, the truth-function $t(A)$ is given explicitly by the well-known truth-value tables, e.g.

$A \wedge B$	1	0	0	0
A	1	1	0	0
B	1	0	1	0

(2.53)

It has been shown by Gleason[18] and Kamber[10] that neither a two-valued truth-function $t(A)$, which satisfies (T_1), (T_2), (T_3), nor a conveniently generalized truth-function, exists on the lattice L_q. Hence, a logical interpretation of this lattice cannot be based on a truth-value semantic. On the other hand, this result does not exclude a logical interpretation of L_q, since it is well known that also the implicative lattice L_i of the intuitionistic logic cannot be interpreted by truth-values[19] In this case, it is the missing principle of excluded middle (tertium non datur) which makes a valuation of L_i by a two-valued truth-function impossible.

However, an implicative lattice L_i can be considered as a logical calculus if one takes into account a more general semantic, the operational interpretation which makes use of the dialogic technique. Propositions $A \in L_i$, which can be shown to be true (i.e. for which the relation $V \leqslant A$ can be derived in L_i) are just those compound propositions for which there exists a strategy of success within a dialog-game, the rules of which will be formulated later. This means that a conveniently defined dialog-game can be considered as a logical semantic of the lattice L_i. However, this dialogic semantic cannot directly be applied to the lattice L_q. The first reason is that this semantic does not lead to an interpretation of the *tertium non datur* (i.e. of the relation $V \leqslant A \vee \neg A$), which is valid in L_q. On the other hand, the dialogic method allows for the justification of some relations (e.g. $A \leqslant B \rightarrow A$), which are not generally valid in L_q.

Although, for the above reasons, this interpretation cannot be applied directly to the lattice L_q, it turns out that a generalization of the dialogic method can be used as an interpretation of a lattice L_{qi} which may be considered as the intuitionistic part of L_q; i.e. if one adds to the axioms of L_{qi} the 'tertium non datur' $V \leqslant A \vee \neg A$, the resulting lattice is isomorphic to L_q. Finally, it can be shown that the dialogic interpretation can be extended also to the lattice L_q. In order to demonstrate the possibility of a logical interpretation of the lattices L_{qi} and L_q, we first formulate the generalized quantum dialog-game (Chapters 3 and 4) which leads to an interpretation of the 'quasi-implicative' lattice L_{qi} (Chapter 5). As a final step, we investigate the extension of this dialogic semantic to the lattice L_q (Chapter 6).

NOTES AND REFERENCES

[1] As a standard reference on lattice theory, we mention here G. Birkhoff, *Lattice Theory*, third edn., American Mathematical Society, Providence, Rhode Island (1973).

[2] In the literature, the lattice L_q is often called *orthomodular*. Here we avoid this terminology, since L_q is orthocomplemented but not *modular*, and thus the expression *orthomodular* is somewhat misleading.

[3] G. Birkhoff, *Lattice Theory*, loc. cit., p. 44.

[4] The question whether L_q is uniquely complemented has been subject of a controversial discussion between Popper[5], Jauch and Piron[6] and others. The entire debate has been reviewed by M. Jammer[7] and critically analyzed by E. Scheibe[8].

[5] K.R. Popper, *Nature* **219** (1968) 682.

[6] J.M. Jauch and C. Piron, in *Quanta*, P.G.O. Freund et al., (Eds). University of Chicago Press, Chicago (1970) p. 166.

[7] M. Jammer, *The Philosophy of Quantum Mechanics*, John Wiley and Sons, New York (1974) p. 351ff.

[8] E. Scheibe, *Br. J. Philos. Sci.* **25** (1974) 319.

[9] M. Nakamura, *Kodai Math. Series*, Rep. **9** (1957) 158.

[10] F. Kamber, *Math. Ann.* **158** (1965) 158.
[11] S. Holland, in: S.C. Abbott, (Ed.), *Trends in Lattice Theory*, Van Nostrand Rheinhold, New York (1970) p. 41ff.
[12] D. Foulis, *Portugaliae Mathematica* **21** (1962) 65.
[13] F. Kamber (ref. 10); also *Nach. Akad. Wiss.* Math. Phys. Klasse **10**, Göttingen, (1964) p. 103; English translation in: C.A. Hooker, (Ed.) *The Logico-Algebraic Approach to Quantum Mechanics* I, D. Reidel Publishing Co., Dordrecht Holland (1975) p. 221.
[14] C.f. H.B. Curry, *Foundations of Mathematical Logic*, McGraw-Hill, New York (1963) in particular Chapter 5, "The theory of implication".
[15] P. Mittelstaedt, *Z. Naturforsch.* **27a** (1972) 1358.
[16] In an implicative lattice L_i – the axioms of which are discussed in the next section – these two relations are equivalent but not generally valid. If they are valid and if a zero element exists, L_i is isomorphic to a Boolean lattice.
[17] H.B. Curry, (ref. 14) uses the term '*implicative*' whereas G. Birkhoff, *Lattice Theory*, loc. cit., p. 45, calls these lattices *relatively pseudo-complemented* or *Browerian.*
[18] A.M. Gleason, *J. Math. Mech.* **6** (1957) 885.
[19] P. Lorenzen, *Formal Logic*, Reidel Publishing Co., Dordrecht, Holland (1965).

CHAPTER 3

THE MATERIAL PROPOSITIONS OF QUANTUM PHYSICS

In this chapter, we develop some basic structures of a language of quantum physics. Starting from elementary propositions about measuring results in Section 3.1, the concept of a dialog is introduced which serves as the most general frame for argumentation. In Section 3.2, we further specify the dialog-game by defining several kinds of compound propositions by argument-rules which prescribe the proof procedures for the compound propositions. In Section 3.3, the fundamental concept of commensurability is introduced by purely operational means and without any recourse to lattice theory. Using this commensurability concept the argument-rules can be reformulated in an adequate manner and the material dialog-game can be established in its final version (Section 3.4).

3.1 ELEMENTS OF A LANGUAGE OF QUANTUM PHYSICS

In this chapter, we shall establish a scientific language of quantum physics which starts from elementary propositions concerning measuring results and which does not take account of the formal results of Chapters 1 and 2. In order to define more complex structures in this language, the concept of dialog is introduced on a purely pragmatic level. Several kinds of compound propositions will then be defined by the possibilities of proving or disproving these propositions within the framework of a dialog. It should be emphasized that for the construction of the formal language of quantum physics, no empirical results will be used. On the contrary, it is this language which makes it possible to speak about experience at all; i.e. the syntactic rules of this language are rather the linguistic preconditions of the possibility of experience. The prescribed procedures for proving or disproving propositions about the real world merely determine the syntactical structure of our scientific language. Hence, these structures may be considered as cognitions which are a priori valid. This is the case, in particular, for the *logical* structure which is contained in the formal language considered here, and will be called *quantum logic*.

3.1.1 *Elementary Propositions*

We start from a quantum mechanical system S (atom, nucleus, elementary particle), the properties of which can be tested by experiment. Here, the physical system S is introduced as a primitive concept which will not be further explained.[1] A proposition $a = a(S, t)$ which asserts that the system S at the time t has a certain measurable property will be called an *elementary proposition*. Elementary propositions will be denoted by $a, b, c, \ldots$. We shall assume here that an experimental procedure M is known which can be considered as a *proof* of the proposition a, or another procedure $\bar{M}$ which can be considered as a *disproof*. Propositions which have this property are said to be *proof-definite* or *disproof-definite*, respectively. We use the following notation:

$M \vdash a \rightleftharpoons$ proposition a has been proved by M,
$\bar{M} \vdash\!\!\!\top a \rightleftharpoons$ proposition a has been disproved by $\bar{M}$.

The concepts of proof and disproof must be defined such that there is no system S for which the proposition considered can be proved and disproved at the same time.

For a proposition a which is disproof-definite, a proof-definite *counter-proposition* $\bar{a}$ can be defined such that

$$\vdash\!\!\!\top a \curvearrowleft\!\!\curvearrowright \vdash \bar{a}. \tag{3.1}$$

Conversely, for a proof-definite proposition, a disproof-definite counter-proposition $\bar{a}$ can be defined by

$$\vdash a \curvearrowleft\!\!\curvearrowright \vdash\!\!\!\top \bar{a}. \tag{3.2}$$

For a proposition a which is proof-definite as well as disproof-definite, the counter-proposition $\bar{a}$ fulfills both conditions (3.1) and (3.2). In this case, we have

$$\vdash \bar{\bar{a}} \leftrightarrow \vdash\!\!\!\top \bar{a} \leftrightarrow \vdash a \tag{3.3}$$

and the propositions $\bar{\bar{a}}$ and a are in this sense equivalent.

In many cases, the propositions a and $\bar{a}$ have the additional property, that for every system an experimental test gives a well-defined result; that is, $\vdash a$ or $\vdash \bar{a}$. Propositions which have this property will be called *value-definite. In the following, we shall assume that elementary propositions are always value-definite.* It is obvious that a value-definite proposition is proof-definite as well as disproof-definite.

As an example of a value-definite proposition, we consider the proposition a which corresponds to the projection operator P_a with eigenvalues 0 to 1. If φ is the state of the system, we have the relations

$$
\begin{aligned}
&\vdash a \leftrightarrow P_a\varphi = \varphi \leftrightarrow (1 - P_a)\varphi = 0 \leftrightarrow \Vdash \bar{a}\\
&\vdash \bar{a} \leftrightarrow P_{\bar{a}}\varphi = \varphi \leftrightarrow (1 - P_{\bar{a}})\varphi = 0 \leftrightarrow \Vdash a
\end{aligned}
\tag{3.4}
$$

which show that the counter-proposition $\bar{a}$ corresponds to the projection operator $P_{\bar{a}} = 1 - P_a$. Since the projection operators P_a and $P_{\bar{a}}$ have only the eigenvalues 0 and 1, any measuring process will lead to one of these values; i.e. either to $\vdash a$ and $\Vdash \bar{a}$ or to $\vdash \bar{a}$ and $\Vdash a$. This is, however, only a special example and we will not assume here that elementary propositions generally correspond to projection operators.

3.1.2 *The Concept of a Dialog*

Elementary propositions are defined by the experimental procedures which serve as a means to prove or to disprove the respective proposition. In the next step, we define compound propositions by means of dialogs. Dialogs serve as proof procedures for compound propositions, whereas the elementary propositions of which they consist must be tested by experiment. A dialog is a formalized kind of discussion between two participants, the proponent P, who asserts a certain proposition about a system S, and the opponent O, who attempts to refute this proposition. The proof or disproof of a compound proposition about S is thus partly the result of a dispute and partly the result of measuring processes. Hence, the relations among P, O and S can be illustrated by the scheme of Fig. 3.1 which shows that P and O are the active participants in the dialog whereas the system S plays merely a passive role.[2]

Without specifying how arguments are to be handled in the course of the dialog the *concept of a dialog* can be defined operationally by some rules which explain how a dialog should be carried out and which will be called *frame-rules*:[3]

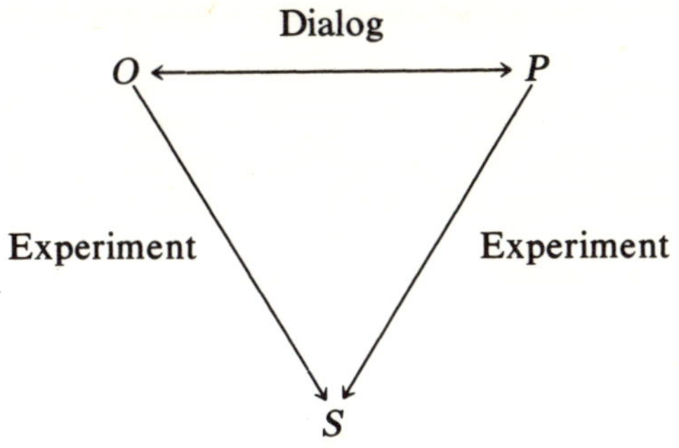

Fig. 3.1 The relations between P, O and S in a dialog.

F1: At the beginning of the dialog the proponent P asserts the initial argument. In this way, the initial position of the dialog is established.

F2: The opponent O attempts to refute this assertion. The dialog then consists of a series of arguments which are stated in turn by the two participants P and O, and which obey certain rules.

F3: Arguments are either attacks on or defences of previous arguments, but not both.

F4: (a) The participants have the right to invoke an attack at any position of a dialog.
(b) Having been attacked, the participants are obliged to defend in the reverse succession of the respective attacks, at the latest when there is no opportunity of attack left; i.e. the latest obligation has to be performed first.

F5: If one of the participants has no argument to continue, he loses the dialog. In this case, the other one wins and the final position of the dialog is established.

In the framework of the dialog defined by (F1)–(F5), arguments are either attacks or defences of certain propositions. The manner in which a proposition can be defended or attacked depends on the possibilities of proving or disproving this proposition. These possibilities will be formulated for various types of propositions by means of *argument-rules* which specify and complete the frame rules of the dialog. For elementary propositions $a, b, \ldots \bar{a}, \bar{b}, \ldots$ the possibilities of proving or disproving have already been mentioned and can be summarized in the following table:

Elementary propositions	Possibility of attack	Possibility of defence
a	$a?$	$a!$
$\bar{a}$	$\bar{a}?$	$\bar{a}!$

By $a?$ we denote the challenge to prove a, by $a!$ the successful proof of a. If a is not only proof-definite but disproof-definite as well, then the counter proposition $\bar{a}$ is also proof-definite, and the proof of $\bar{a}$ is given by the disproof of a. In the above attack-and-defence scheme, it has been assumed that this is actually the case.

It is obvious that the proof procedure for an elementary proposition a can also be expressed in terms of dialogs, although the notion of a dialog was not conceived for this special case. A dialog in which the proponent P successfully defends an elementary proposition a and which takes account of the above-mentioned possibilities of attack and defence can then be written in the following way:

	O	P
(0)	[ ]	a
(1)	$a?\,(0)$	$a!\,\langle 0\rangle$

Here we have used a schematic representation of the dialog which will frequently be applied in the following chapters. Hence, it is useful at this point to formulate the rules of this representation in a complete form.

Schematic Representation of Dialogs

(1) The dialog is represented by means of two columns. The left column consists of all arguments of O, the right column of all arguments of P.

(2) The succession of arguments is enumerated by rows which begin at zero. The argument of P in Row zero is the initial argument, whereas the argument of O in Row zero is the empty argument.

(3) If an argument is an attack against a statement which has been asserted in Row x, we write (x) on the right-hand side of the argument.

(4) If an argument is a defence of an assertion which has been

stated in Row y and attacked later in the dialog, we write $\langle y \rangle$ on the right-hand side of the defending argument.

(5) If a participant does not defend, the empty argument [ ] is placed in the respective row, and he continues in the next row.

These rules are partly illustrated by the dialog (3.5). The proponent asserts in $P(0)$ the proposition a, and the opponent attacks this statement in $O(1)$ by a?. In $P(1)$, the proponent defends his assertion by a!, which completes the dialogic proof.

If a dialog-game D_x has once been established by convenient argument-rules, it is useful to introduce the following terminology: A participant (P or O) is said to possess a *strategy of success* within the dialog-game D_x if he knows a succession of arguments by means of which he wins the dialog with certainty irrespective of the arguments of the other participant. Using the concept of dialogic strategy, we can define the following important notions.

(3.6) DEFINITION: A proposition A is said to be *true* (or x-true) if and only if P has a strategy of success within the dialog-game D_x *about* A; i.e. if P wins the dialog irrespective of the arguments of O.

(3.7) DEFINITION: A proposition A is said to be *false* (or x-false) if and only if P has a strategy of success *against* A; i.e. if O asserts A in a dialog, P wins the dialog within the game D_x irrespective of the arguments of O.

By means of these definitions every dialog-game D_x, which is characterized by certain argument-rules, is related to corresponding concepts of truth and falsity. If a proposition A is true with respect to the game D_x we write

(3.8) DEFINITION: $\vdash_{D_x} A \rightleftharpoons$ The proposition A is true with respect to the dialog-game D_x.

3.2 ARGUMENT-RULES FOR COMPOUND PROPOSITIONS

Elementary propositions are proof-definite and disproof-definite, respectively. Propositions which are composed of elementary propositions are defined by the possibilities of proving or disproving these propositions by means of a dialog (i.e. by the respective *argument-rule*). Hence, *compound propositions* are said to be *dialog-definite.* Similarly, as for elementary propositions, the argument-rule for a certain compound proposition consists of a specification of all possibilities of attack and of defence of this proposition. In the following, we shall consider two types of compound propositions, the *sequential propositions*, which are time-dependent compositions of elementary propositions, and time-independent compound propositions, which correspond to the *logical connectives.*

3.2.1 *Sequential Propositions*

Elementary propositions $a = a(S, t)$ are, in general, time-dependent. Starting from several time-dependent propositions $a(S, t_1)$, $b(S, t_2)$, $c(S, t_3), \ldots$ which are related to different time values $t_1 \leqslant t_2 \leqslant t_3$, one can define sequences $\sigma\{a, b, c, \ldots\}$ of propositions which depend on the temporal order of the propositions $a, b, c \ldots.$ Here we are considering only 2-place sequences (i.e. expressions of the form $\sigma\{a, b\}$, with $t(a) \leqslant t(b)$). There are at least four different kinds of 2-place sequences, which we define by the possibilities of attacking and defending these propositions in a dialog. These possibilities (i.e. the argument-rules) are summarized in the table below.

We have added here the *colloquial denotation* '*a* and then *b*', '*a* or then *b*', 'if first *a* then *b*', and 'not *a*' in order to give at least a vague idea of the empirical meaning of these sequences. They have,

sequence $\sigma\{a, b\}$	colloquial denotation	possibility of attack	possibility of defence
$a \sqcap b$	*a* and then *b*	1? 2?	*a* *b*
$a \sqcup b$	*a* or then *b*	?	*a* *b*
$a \dashv b$	if at first *a* then *b*	*a*	*b*
$\urcorner a$	not *a*	*a*	

however, no influence on the use of the sequences in a dialog, which is governed exclusively by the argument-rule formulated in the above table. As an illustration of the dialogic meaning of these formal definitions, let us consider the proposition $a \dashv b$, or more precisely, $a(S, t_1) \dashv b(S, t_2)$. If the proponent P, say, asserts the proposition $a \dashv b$, then he assumes the obligation, in the case that the opponent O can prove $a(S, t_1)$ by an experiment, to prove $b(S, t_2)$ by another experiment at the later time $t_2 > t_1$.

It is obvious from this explanation of sequences that in a time-dependent dialog an attack or a defence can be performed only once (i.e. at the respective time values). Since a 2-place sequence $\sigma\{a(S, t_1), b(S, t_2)\}$ is related to two different time values t_1 and t_2, it follows that the respective dialog consists essentially of two steps which correspond to two separate measuring processes. Thus, the schematic representation of the dialogs for sequential propositions read e.g.:

	O	P	
(0)	[ ]	$a \sqcap b$	
(1)	1? (0)	a ⟨0⟩	
(2)	a? (1)	a!⟨1⟩	t_1
(3)	2? (0)	b ⟨0⟩	
(4)	b? (3)	b!⟨3⟩	↓ t_2

	O		P		
(0)	[ ]		$a \sqcup b$		
(1)	?		a		t
(2)	a?		a!	[ ]	
(3)				b	
(4)		b?		b!	↓

(3.9)

	O	P	
(0)	[ ]	$a \dashv b$	
(1)	a (0)	[ ]	
(2)	a!⟨1⟩	a? (1)	t_1
(3)	[ ]	b ⟨0⟩	
(4)	b? (3)	b!⟨3⟩	↓ t_2

	O	P	
(0)	[ ]	$\neg a$	
(1)	a (0)	[ ]	
(2)	a!⟨1⟩	a?	t_1
			↓ t

Here we have assumed that P, in fact, wins the dialogs for $a \sqcap b$, $a \sqcup b$, $a \dashv b$ and that he loses the dialog about $\neg a$. In order to indicate the temporal order of the different steps in the dialog, we have added here a t-axis, the metric of which is irrelevant.

We are not going to investigate in detail the consequences of these definitions and the syntactic structure which can be obtained from them. In particular, we shall not formulate here a calculus of sequences[7] and the corresponding abstract algebra.[8] Instead, we shall use sequential propositions merely as a starting point for establishing those compound propositions which are definitely time-independent, and which are usually associated with logical connectives. In this way, it will become clear that the time-independence of logical statements is very nearly related to the problem of commensurability,[9] which will be discussed in the next section.

3.2.2 *Logical Connectives*

Sequential propositions $\sigma\{a(S, t_1), b(S, t_2)\}$ are related to two different time values t_1 and t_2. In a similar way, the *logical connectives* are defined in terms of 2-place compound propositions $\lambda\{a(S, t), b(S, t)\}$, the elementary subpropositions of which are related to the system S at the *same time value* t. Just as in the case of sequential propositions, we define these *simultaneous* compositions of elementary propositions by the possibilities of attacking and defending them in a dialog. In order to avoid any confusion with the sequential propositions for the logical connectives, we use the traditional names and a slightly different denotation. Again, we define four compound propositions which can be shown to be a minimal basis for all logical connectives:

connective	symbol	colloquial denotation	possibility of attack	possibility of defence
conjunction	$a \wedge b$	a and b	1? 2?	a b
disjunction	$a \vee b$	a or b	?	a b
material implication	$a \rightarrow b$	a then b	a	b
negation	$\neg a$	not a	a	

Again the colloquial denotations 'a and b', 'a or b', 'a then b' and 'not a' have been added here only for mnemonic support. As in the case of the sequential propositions, they are without any influence on the use of the connectives in a dialog. At first glance, the argument-rules for the sequential propositions and the logical connectives seem to be equivalent. In fact, in both cases we have the same possibilities of attack and defence. There is, however, a decisive difference which stems from the fact that the elementary subpropositions of a logically connected proposition are related to the same time value. For the dialogic proof of a sequential proposition $\sigma\{a(S, t_1), b(S, t_2)\}$ with $t_1 < t_2$, there are at least two measuring processes to be performed at time values t_1 and t_2, respectively. No attack or defence can be repeated. On the other hand, in the limit of a logically connected proposition the two elementary propositions are simultaneously related to the system S. Hence there is no limitation on the number of attacks and defences for the elementary propositions in a dialog. For this reason, the dialog about a logical connective will, in principle, be of infinite length. This can easily be illustrated by the following examples.

(α) *Conjunction* $a \wedge b$: According to the argument-rule, the proponent, who has asserted a conjunction $a \wedge b$, is committed to demonstrating the elementary propositions $a(S, t)$ and $b(S, t)$ by experimental tests. We will assume that these proofs have been performed in $P(2)$ and $P(4)$ (see (3.10)). Since the propositions a and

	O	P
(0)	[ ]	$a \wedge b$
(1)	1? (0)	$a \langle 0 \rangle$
(2)	a? (1)	$a! \langle 1 \rangle$
(3)	2? (0)	$b \langle 0 \rangle$
(4)	b? (3)	$b! \langle 3 \rangle$
(5)	1? (0)	$a \langle 0 \rangle$
(6)	a? (5)	

(3.10)

b are related to the same time t, the opponent can attack proposition a again, as in Row $O(2)$. This is indicated here in Row $O(6)$. Hence there is no limitation on the length of the dialog. One could dispense with the repetition of the experimental proof for a by the proponent in $P(6)$ *only* if there is some guarantee that the result $a!$ of the a-measurement in $P(2)$ would still be valid – even after the b-test has been performed in $P(4)$ – and therefore still *available* for the proponent.

(β) *Disjunction* $a \vee b$: For the discussion of the dialog about the disjunction $a \vee b$, we assume that the proponent did not succeed in proving the propositions a and b by experiment in $P(2)$ and $P(4)$, which is indicated in (3.11) by $\not a!$ and $\not b!$, respectively. Since a and b

	O	*P*
(0)	[ ]	$a \vee b$
(1)	? (0)	a ⟨0⟩
(2)	a? (1)	$\not a$! ⟨1⟩
(3)	[ ]	b ⟨0⟩
(4)	b? (3)	$\not b$! ⟨3⟩
(5)	[ ]	a ⟨0⟩
(6)	a?	a! ⟨5⟩

(3.11)

are again simultaneously related to the system S, the dialog is not yet definitely lost for P in Row 4. For it could happen that after the experimental test of b in $P(4)$ a new measurement of the proposition $a(S, t)$ in $P(6)$ leads to the desired result $a!$, even if the a-test in $P(2)$ was unsuccessful. Hence the opponent is committed to attack again in $O(6)$ and, in this way, the dialog can continue to infinite length. The possibility of a successful experimental test in $P(6)$ could only be excluded if there were some guarantee that the negative result of the a-test in $P(2)$ would be reobtained in a new experiment, even after the b-measurement in $P(4)$.

(γ) *Material implication* $a \rightarrow b$: If the proponent asserts the proposition $a \rightarrow b$, then he assumes the obligation, in the case that $a(S, t)$ can be proved by the opponent, of justifying b by another experiment. Here, we have assumed that the proponent P can actually perform the proof of b in $P(4)$ (see (3.12)). Since the propositions a

	O	*P*
(0)	[ ]	$a \rightarrow b$
(1)	a (0)	[ ]
(2)	a! ⟨1⟩	a? (1)
(3)	[ ]	b ⟨0⟩
(4)	b? (3)	b! ⟨3⟩
(5)	a (0)	[ ]
(6)	a! ⟨5⟩	a? (5)
(7)	⋮	⋮

(3.12)

and b are related to the same time, the opponent could attack again by asserting a, as in Row $O(1)$. Hence, there is no limitation on the dialog. One could dispense with the repetition of the experimental proof of b by P only if there were some guarantee that the result $b!$ of the b-measurement in $P(4)$ would still be valid after the new a-attack of the opponent (and the respective experiment in $O(6)$), and, therefore, still available for the proponent.

(δ) *Negation* $\neg a$: In contrast to the three 2-place connectives, the dialog about the negation can be made finite without further assumptions. Since the negation contains only one elementary proposition $a(S, t_1)$ which is related to a single time value t_1, there is no difference between the negation and the respective sequential proposition. (For this reason, we have used the same denotation in both cases.) Here, we have assumed that the opponent has given a proof of $a(S, t_1)$ in Row $O(2)$ (see (3.13)). In this situation, the proponent has definitely

	O	*P*
(0)	[ ]	$\neg a$
(1)	a (0)	[ ]
(2)	$a!\langle 1\rangle$	$a?(1)$
(3)	$a!\langle 1\rangle$	$a?(1)$

(3.13)

lost the dialog. In fact, if the proponent were to continue the dialog and repeat the attack $a?$ at $P(4)$, the opponent would again successfully defend by proving $a!$ in $O(3)$. Since, between the a-proof in $O(2)$ and its repetition in $O(3)$, no other proposition has been tested experimentally, there is no reason to doubt that in both cases the same result will be obtained. Hence the dialog may be considered as already finished in Row 2.

The logical connectives $\wedge$, $\vee$ $\rightarrow$ and $\neg$ can be used in order to extend the set of propositions which can be treated in a dialog. Starting from the set $S_e = a, b, c \dots$ of elementary propositions we extend this set by incorporating arbitrary iterations of the logical connectives. The resulting set S of propositions is then given by the following inductive scheme:

DEFINITION: The set S of propositions is given by:

(I) Elementary propositions $a \in S_e$ are propositions.

(II) If A and B are propositions, i.e. $A \in S$, $B \in S$, then $A \wedge B$, $A \vee B$, $A \rightarrow B$, $\neg A$ are propositions.

Elements of S will be denoted here by $A, B, C \ldots$. If an element $A \in S$ is an elementary proposition, it has to be proved outside of the dialog by a measuring process. Elementary propositions are *proof-definite.* If $A \in S$ is a compound proposition which contains at least one logical connective, the proof of this proposition has to be performed by a dialog. Hence, compound propositions are said to be *dialog-definite.* The complete proof of a compound proposition A then consists in a dialogical decomposition of A, and in experimental proofs, of the elementary subpropositions of A. Dialogs which contain proof procedures which have to be performed outside of the dialog will be called *material dialogs.*

3.3 COMMENSURABILITY AND INCOMMENSURABILITY

In the preceding section, it has been shown that the logical connectives $a \wedge b$, $a \vee b$ and $a \rightarrow b$ correspond to infinite dialogs, whereas the dialog of the negation $\neg a$ can be restricted to a finite number of steps. The considerations which lead to this result were based on two assumptions α_1, α_2 which have been tacitly presupposed, but which should be explicitly mentioned. The argument – that the dialog concerning the negation may be confined to a finite number of steps – is based on the assumption:

(α_1) If proposition $a(S, t)$ has been proved by measurement, then an immediate repetition of the measurement at time $t' = t + \delta t$ will also result in a, if δt is sufficiently small.

On the other hand, the conclusion that the dialogs for the connectives $\wedge$, $\vee$ and $\rightarrow$ are infinite was based on the following argument:

(α_2) If proposition $a(S, t)$ has been proved by an experiment, then after a measurement of proposition $b(S, t')$ at $t' > t$, in a subsequent measurement (at $t'' > t'$), one does not necessarily obtain $a(S, t'')$, even if the time difference $\delta t = t'' - t$ is arbitrarily small.

The first assumption α_1 is in complete agreement with our daily life experience and the only reason to mention it here explicitly is that, with respect to assumption α_2, one might even doubt the validity of α_1. The second assumption α_2 seems to be more restrictive than is actually necessary. In fact, no experience is known in our daily life which would justify such a cautious assumption. However, we know from quantum mechanics that the validity of a proposition $a(S, t)$ once proved for the system S, is no longer guaranteed if other propositions have been proved later by an experiment in the same system. The theory of measurement in quantum mechanics has shown that due to the incommensurability of measurable properties, the result of the measurement of a certain observable $\mathcal{P}_a$ can be completely destroyed by the measurement of another observable $\mathcal{P}_b$ which is not commensurable with $\mathcal{P}_a$.

However, at the present stage of our discussion, we do not like to refer to any empirical knowledge. We have mentioned these quantum mechanical results here only in order to motivate the precautious assumption α_2, which is obviously suited to incorporate situations such as those which are known from quantum physics. Instead of making any empirical assumption with respect to the question whether a proposition $a(S, t)$, once proved in a dialog, is still *available* for the proponent after the experimental test of another proposition $b(S, t)$, we shall rather incorporate here an additional testing procedure into the rules of the dialog which decides in every case about the *availability* of the respective proposition. This will be done by means of *availability propositions*. In this way, it will be possible to establish a dialog-game which is applicable to propositions about quantum-mechanical systems as well as to propositions about classical systems.[10]

In order to perform this program, we first introduce two '*availability propositions*' $k(A, B)$ and $\bar{k}(A, B)$, which state whether a given proposition A is still available after a material dialog about the proposition B, or not. The availability propositions must be proved or disproved outside the dialog by procedures which follow from the precise definition of the propositions $k(A, B)$ and $\bar{k}(A, B)$.

The propositions $k(A, B)$ and $\bar{k}(A, B)$ will be called *commensurability* and *incommensurability*, respectively, and will be defined in the following way:

(3.14) DEFINITION $k(A, B)$: Two propositions A and B are said to be *commensurable* – $k(A, B)$ has been proved – if, in a given system S, the propositions A and B can be tested dialogically in an arbitrary sequence without thereby influencing the result of the dialogs.

(3.15) DEFINITION $\bar{k}(A, B)$: Two propositions A and B are said to be *incommensurable* – $\bar{k}(A, B)$ has been proved – if, in a given system S, the result of a dialogic test of the one proposition can be changed by a dialogic test of the other proposition.

According to these definitions, the proofs of $k(A, B)$ and $\bar{k}(A, B)$, which are performed outside the dialog, consist of a series of alternating dialogs about A and B (if A and B are elementary propositions, in a series of experimental tests of A and B). $k(A, B)$ is proved if and only if in each arbitrary sequence of dialogs about A and B the dialogs are won by the same participants, respectively. The incommensurability $\bar{k}(A, B)$ is proved if and only if in a sequence of dialogs about A and B the result of a dialog about A or B, respectively, changes. It is obvious that these proof procedures for the availability propositions $k(A, B)$ and $\bar{k}(A, B)$ require an *infinite* number of steps. The consequences of this important fact will, however, be discussed later.

It follows from the definitions of $k(A, B)$ and $\bar{k}(A, B)$ and from the corresponding proof procedures that $k(A, B)$ and $\bar{k}(A, B)$ are counter-propositions, i.e.

$$\vdash k(A, B) \curvearrowright \mid\!\!\top \bar{k}(A, B), \qquad \vdash \bar{k}(A, B) \curvearrowright \mid\!\!\top k(A, B). \tag{3.16}$$

Similarly, as for elementary propositions, the possibilities of proving or disproving propositions $k(A, B)$ and $\bar{k}(A, B)$ in a dialog can be summarized in the following table:

Availability propositions		Possibility of attack	Possibility of defence
Commensurability	$k(A, B)$	$k(A, B)?$	$k(A, B)!$
Incommensurability	$\bar{k}(A, B)$	$\bar{k}(A, B)?$	$\bar{k}(A, B)!$

Since the proofs for elementary propositions and for commensurabilities have to be performed outside the dialog, these two kinds of propositions will be called *material propositions*.

The possibilities of attacking and defending *material propositions* (i.e. $a?$, $k(A, B)?$, $\bar{k}(A, B)?$ and $a!$, $k(A, B)!$, $\bar{k}(A, B)!$, respectively) are not arguments in the strict sense of the frame-rules. They are rather questions and answers which are related to the third 'passive participant' in the dialog-game, the physical system S. In order to incorporate this kind of argument into the dialog-game in an adequate way, the following weakening of the frame-rule F(4b) will be formulated here as an argument rule for material propositions:

A_m: If a participant cannot defend against an attack of a material proposition $(a, k(A, B), \bar{k}(A, B))$ he may assume a previous obligation of defence.

The reason for this liberalisation of the dialog-rules is based on the fact that neither the proponent nor the opponent has any influence on the proof result of a material proposition. Therefore, one should allow the participants P and O to check the truth of several propositions by the respective proof-attempts.

If the propositions $k(A, B)$ and $\bar{k}(A, B)$ have once been defined, the infinite dialogs about the logical connectives $A \wedge B$, $A \vee B$, $A \rightarrow B$ and $\neg A$ can be replaced by dialogs which contain only a finite number of steps. This can easily be illustrated by the following dialogs about the connectives $a \wedge b$, $a \vee b$, $a \rightarrow b$, which contain only elementary propositions.

(α) *Conjunction* $a \wedge b$: The infinite dialog (3.10) about $a \wedge b$ could be limited to a finite length if it were known that a and b are commensurable. This, however, can be tested by a proof-attempt outside the dialog for the commensurability proposition $k(a, b)$. Hence, we extend the possibilities of attacking $a \wedge b$ in a dialog by the additional attack $k(a, b)$? with the respective defence $k(a, b)$!. The dialog (3.10) can then be replaced by the finite dialog (3.17). Here, P is assumed to

	O		P		
(0)	[ ]		$a \wedge b$		
(1)	$k(a, b)?$	(0)	$k(a, b)!$	⟨0⟩	
(2)	1?	(0)	a	⟨0⟩	
(3)	$a?$	(2)	$a!$	⟨2⟩	(3.17)
(4)	2?	(0)	b	⟨0⟩	
(5)	$b?$	(4)	$b!$	⟨4⟩	

succeed in defending against all attacks of O. Since a and b are commensurable, subsequent proofs of a and b will reproduce the same results. Therefore, the dialog (3.17) can be terminated after the successful proof of b in Row $P(5)$. In case a and b are incommensurable (i.e. $k(a, b)$ cannot be proved by P) the proponent loses the dialog (3.17) in Row $P(1)$. In the unlimited dialogic procedure the incommensurability of a and b means that the proponent would not be able to defend a and b in an arbitrarily long series of challenges by O. Hence P is certain to lose the unlimited dialog about $a \wedge b$. In this way, it can be seen that the commensurability proposition in the dialog (3.17) indeed replaces a dialog with unlimited possibilities of attacking by 1? and 2?.

(β) *Disjunction* $a \vee b$: The infinite dialog (3.11) about $a \vee b$ could be limited to a finite length if it were known that a and b are incommensurable. This can again be tested outside the dialog by a proof attempt for the incommensurability proposition $\bar{k}(a, b)$. Therefore we will extend the possibilities of defending $a \vee b$ by the additional defence $\bar{k}(a, b)$ with the respective attack $\bar{k}(a, b)?$. By means of this new possibility, the dialog (3.11) can be replaced by the finite dialog (3.18). As in the dialog (3.11), we have assumed here that the proponent fails in his attempts to defend $a \vee b$ by proving a and b. In Row $P(5)$ of (3.18) he defends by asserting the incommensurability $\bar{k}(a, b)$. If he can prove $\bar{k}(a, b)$, as assumed in $P(6)$, then in a series of further proof attempts for a and b he will succeed in proving a or b, respectively. Therefore, the proponent has a strategy of success in the unlimited dialog also. In case P cannot prove the incommensurability proposition $\bar{k}(a, b)$, he has no strategy of success in the unlimited dialog about $a \vee b$ and thus loses the game. In this way, we see that the incommensurability proposition $\bar{k}(a, b)$ in the dialog (3.18) indeed replaces an unlimited series of assertions of a and b by the proponent.

	O		P	
(0)	[ ]		$a \vee b$	
(1)	?	(0)	a	⟨0⟩
(2)	a?	(1)	$\not{a}$!	⟨1⟩
(3)	[ ]		b	⟨0⟩
(4)	b?	(3)	$\not{b}$!	⟨3⟩
(5)	[ ]		$\bar{k}(a, b)$	⟨0⟩
(6)	$\bar{k}(a, b)$?	(5)	$\bar{k}(a, b)$!	⟨5⟩

(3.18)

(γ) *Material implication* $a \rightarrow b$: The dialog (3.12) about $a \rightarrow b$ could be limited to a finite length if the proponent could guarantee the availability of a in Row $P(6)$. This can, however, be tested by a proof attempt for the commensurability $k(a, b)$. Therefore, we extend the possibilities of attacking $a \rightarrow b$ in a dialog by the additional attack $k(a, b)$? and the respective defence $k(a, b)$!. The infinite dialog (3.12) can then be replaced by the finite dialog scheme (3.19). Before

	O		P	
(0)	[ ]		$a \rightarrow b$	
(1)	a	(0)	[ ]	
(2)	$a!$	⟨1⟩	$a?$	(1)
(3)	[ ]		b	⟨0⟩
(4)	$k(a, b)?$	(0)	$k(a, b)!$	⟨0⟩
(5)	$b?$	(3)	$b!$	⟨3⟩

(3.19)

challenging b the opponent asks for $k(a, b)$, which, we assume here, is proved by the proponent in $P(4)$. After the defence $b!$ in $P(5)$ the dialogic proof of $a \rightarrow b$ can be considered complete, since according to the proof of $k(a, b)$ the result $b!$ in Row $P(5)$ will be available in the further course of the dialog, even if the opponent attacks again proving proposition a – as in the unlimited dialog (3.12). If P does not succeed in demonstrating $k(a, b)$ he loses the dialog (3.19). This corresponds again to the situation in the unrestricted procedure, where, in this case, P is certain to fail in a proof of b after a sufficiently long series of attacks by a.

The possibility of replacing the infinite dialogs about the logical connectives by finite dialogs with availability propositions is, of course, not restricted to connected elementary propositions. The above considerations can be transferred almost literally to connectives of the more general kind $A \wedge B$, $A \vee B$ and $A \rightarrow B$, where A and B are arbitrary compound propositions. In this case, the material proofs by measurements have simply to be replaced by material subdialogs about the respective compound propositions.

3.4 THE MATERIAL DIALOG-GAME

Elementary propositions $a, b, c \ldots$; $\bar{a}, \bar{b}, \bar{c}, \ldots$ and availability propositions $k(a, b)$ and $\bar{k}(a, b)$ are *proof-definite material propositions*, the proof of which has to be performed outside the dialogs. Starting from the set S_e of elementary propositions in Section 3.2, we have extended S_e by incorporating arbitrary iterations of the logical connectives. The resulting set S has then been defined by the following inductive scheme:

$S=\{A\}$: (I) Elementary propositions $a \in S_e$ are propositions.

(II) If A and B are propositions (i.e. $A \in S$, $B \in S$) then $A \wedge B$, $A \vee B$, $A \rightarrow B$, $\neg A$ are propositions.

From an algebraic point of view, the logical connectives $\wedge$, $\vee$, $\rightarrow$ and $\neg$ are 1- and 2-place operations on the set S. In a second step, we extend the set S by incorporating also the availability propositions $k(A, B)$, $\bar{k}(A, B)$ and its iterations. The extended set S^* is then given by the inductive scheme:

$S^*=\{A\}$: (I) Elementary propositions $a \in S_e$ are propositions.

(II) If A and B are propositions (i.e. $A \in S^*$, $B \in S^*$) then $A \wedge B$, $A \vee B$, $A \rightarrow B$, $\neg A$, $k(A, B)$ and $\bar{k}(A, B)$ are propositions.

Apart from the logical connectives on the set S^* the commensurabilities $k(A, B)$ and $\bar{k}(A, B)$ are also 2-place operations on the set S^*. In spite of this algebraic similarity between the logical connectives and the commensurabilities, there is, however, a decisive difference which should be kept in mind. The connectives are dialog-definite propositions, whereas the commensurabilities $k(A, B)$, $\bar{k}(A, B)$ are always proof-definite material propositions – even if the subpropositions A and B, which they contain, are dialog-definite compound propositions.

The complete proof of an arbitrary proposition $A \in S$ or $A \in S^*$ consists, firstly, in a dialogic decomposition of the compound proposition A and, secondly, in material proofs of the material subpropositions of A. These complex proof-procedures for propositions $A \in S$ or $A \in S^*$ will be called the *material dialogic-game* D_m or D_m^*, respectively. Since D_m as well as D_m^* incorporate material proofs for availability propositions $k(A, B)$ and $\bar{k}(A, B)$, the game will also be called more precisely *material quantum dialog-game*. There is an important conceptual difference between S and S^* or D_m and D_m^*, respectively. In the material dialog-game D_m the commensurabilities $k(A, B)$ and $\bar{k}(A, B)$ are only contained insofar as they are availability propositions which serve as a means to restrict the dialogs about the connectives to finite length. In the game D_m^*, the commensurabilities are also contained as independent proof-definite propositions about the physical system in question.

We are now going to investigate the details of the material quantum dialog-game. Within the general frame rules (F1)–(F5) of the dialog, material dialogs are determined by the possibilities of proving or disproving propositions $A \in S$ or $A \in S^*$. These possibilities will be formulated here as *argument-rules* of the material (quantum) dialog-game. In the following, we will formulate the argument-rules with respect to the extended set S^*. The rules of the dialog-game D_m can then easily be obtained by eliminating those commensurabilities $k(A, B)$ and $\bar{k}(A, B)$ which appear as independent propositions. For the formulation of the formal (effective) quantum logic (Chapters 4 and 5) it will turn out to be sufficient to consider only propositions $A \in S$ and the dialog-game D_m. However, for establishing the formalism of full quantum logic (Chapter 6) the extended material dialog-game D_m^* will be indispensable.

The argument rules of the material dialog game D_m^* will be denoted by $A_m(1), \ldots$. These rules summarize the results of the preceding sections in a systematic form: In $A_m(1)$, $A_m(2)$ and $A_m(3)$ the possibilities of proving material propositions are formulated ($A_m(3)$ is identical with the argument rule A_m formulated in the last section). In $A_m(1)$ the counter-proposition $\bar{a}$, which belongs to an elementary proposition, is no longer mentioned. The reason is that within the material dialog-game, and for a value definite proposition a, $\bar{a}$ can be completely replaced by the negation $\neg a$. The proof of this *dialog-equivalence* requires a detailed discussion of the dialogic meaning of the value-definiteness and will thus be given in Chapter 6. The rules A_m(4a–4d) are concerned with the possibilities of proving the logical connectives by means of the availability propositions $k(A, B)$ and $\bar{k}(A, B)$. The reason for this generalisation of the usual definition of the connectives is the potential incommensurability of quantum mechanical propositions, which must be taken into account even in the definition of the logical connectives.

The essential difference between the argument rules for quantum mechanical propositions and the respective rules for propositions of classical physics is the *restricted availability* of the propositions expressed in the rule $A_m(5)$. According to this rule, the availability of a proposition A, which has been asserted at an earlier stage of the dialog, must always be checked by testing the commensurabilities $k(A, B_i)$ $(i = 1, \ldots n)$ of this proposition A and all propositions B_i which have been asserted later in the dialog. The reason for this restriction is obvious: In case A and B_k are incommensurable, the result of a successful proof of A, which consists of a series of measurements, could be completely destroyed by measurements which belong to the proof of B_k.

In the argument-rules of the material (quantum) dialog-game, no special empirical assumptions have been made concerning the commensurability of propositions. Instead, we have rather incorporated into the dialog-rules an additional testing procedure which decides in every case about the availability of a proposition. For that reason, the material dialog-game for quantum mechanical propositions presented here is not restricted to propositions about quantum mechanical systems but can be applied as well to propositions of classical physics.[12]

The Argument Rules of the Material Quantum Dialog-Game D_m

$A_m(1)$: If an elementary proposition a is asserted it may be attacked by a challenge (a?) to prove it. The defence consists in a demonstration of a. (For a successful proof, we write a!) The attack and defence scheme of elementary propositions reads:

Elementary propositions	attack	defence
a	$a?$	$a!$

$A_m(2)$: (a) If a commensurability proposition $k(A, B)$ is asserted it may be attacked by the challenge $k(A, B)$? to prove it. The defence consists in a demonstration of $k(A, B)$. (For a successful proof we write $k(A, B)$!)

(b) An incommensurability proposition $\bar{k}(A, B)$ may be attacked by the challenge $\bar{k}(A, B)$? to prove it. The defence consists in a demonstration of $\bar{k}(A, B)$, which is denoted by $\bar{k}(A, B)$!. The attack and defence scheme of the availability propositions reads:

Availability proposition	attack	defence
$k(A, B)$	$k(A, B)?$	$k(A, B)!$
$\bar{k}(A, B)$	$\bar{k}(A, B)?$	$\bar{k}(A, B)!$

$A_m(3)$: If a participant cannot defend against an attack of a material proposition (i.e. a, $k(A, B)$, $\bar{k}(A, B)$) he may assume a previous obligation of defence.

$A_m(4)$: (a) If a conjunction $A \wedge B$ is stated, it may be attacked once by each of the arguments $k(A, B)?$, 1? and 2?. The corresponding defences are the demonstration of the commensurability of A and B (i.e. $k(A, B)!$), the assertion of A and the assertion of B.

(b) If a disjunction $A \vee B$ is stated it may be attacked by a challenge (?) to defend. The defence consists in the assertion of $\bar{k}(A, B)$, A or B which may be used once each.

(c) If a material implication $A \rightarrow B$ is stated, it may be attacked by the assertion of A. Thereupon, either a defence by B or an attack on A are possible. After B is stated, the commensurability of A and B may be attacked (by $k(A, B)?$). The defence on this attack is a demonstration of the commensurability of A and B ($k(A, B)!$). The next argument may be an attack against B.

(d) If a negation $\neg A$ is stated, it may be attacked by the assertion of A. Thereupon no defence, but only an attack against A, is possible.

The attack and defence scheme which corresponds to this argument-rule is the following:

connective	attacks	defences
$A \wedge B$	$k(A, B)?$ 1? 2?	$k(A, B)!$ A B
$A \vee B$	?	$\bar{k}(A, B)$ A B
$A \rightarrow B$	A $k(A, B)?$	B $k(A, B)!$
$\neg A$	A	

$A_m(5)$: If a participant attacks a proposition A the other participant may check the availability of A by using the arguments $k(A, B)?$. B is any proposition asserted in the dialog after A, and the attack $k(A, B)$? may be used once for every B.

By means of the argument-rules $A_m(1)$–$A_m(5)$ material dialogs can be performed in order to prove or disprove compound propositions $A \in S^*$. According to the terminology introduced above, a proposition $A \in S^*$ which can be successfully defended within D_m^* will be called *materially true* (m-true). Conversely, if the proponent has a strategy of success against A within the dialog-game D_m^*, A will be called *materially false* (m-false). On account of the argument-rule $A_m(4d)$ this is the case if and only if P has a strategy of success for $\neg A$. Hence, we arrive at the following terminology

(3.20) DEFINITION: $\vdash_{D_m^*} A \rightleftharpoons A$ is *materially true*; i.e. P has a strategy of success in D_m^* for A.

$\vdash_{D_m^*} \neg A \rightleftharpoons A$ is *materially false*; i.e. P has a strategy of success in D_m^* for $\neg A$.

An analogous terminology can, of course, be introduced for the set S and the material dialog-game D_m, respectively.

NOTES AND REFERENCES

[1] In many cases, primitive concepts can be eliminated within the framework of the complete theory by theoretical terms only. For the concept of a *physical system in quantum mechanics* this was demonstrated by C. Piron, Varenna lectures, 1977.

[2] In most scientific languages, the object S is not mentioned explicitly. However, we shall find that the possibility of eliminating the object system in the formal language depends on some conditions which are not generally fulfilled.

[3] In the framework of ordinary logic, which is concerned with the special situation of propositions with unrestricted availability (for the notion of availability cf. Section 3.3), the concept of dialog has been treated by P. Lorenzen[4], W. Kamlah and P. Lorenzen[5] and K. Lorenz[6]. However, as long as only the concept of dialog (i.e. its frame rules) are concerned, there is no difference between the investigations of these authors and the present approach.

[4] P. Lorenzen, *Metamathematik*, Bibliographisches Institut, Mannheim (1962).

[5] W. Kamlah and P. Lorenzen, *Logische Propädeutik*, Bibliographisches Institut, Mannheim (1973).

[6] K. Lorenz, *Arch. f. Math. Logik und Grundlagenforsch.* **11** (1968).

[7] The calculus of sequences has been studied in great detail by E.W. Stachow (forthcoming).

[8] An algebraic structure, which is very nearly related to the calculus of sequential propositions, has been developed by H. Kröger, *Sitzungsber. Bayer. Akad. Wiss.*, München, (1973).

[9] The relation between time-independent (quantum) logic and time-dependent elementary propositions has been investigated in detail in: P. Mittelstaedt, in: Proceedings of the Symposium on Quantum Logic, Bad Homburg, Germany (1976), *J. Philos. Logic* **6** (1977) 463.

[10] A detailed comparison of the different situations in classical physics and quantum physics can be found in ref. 11, Chap. VI.

[11] P. Mittelstaedt, *Philosophical Problems of Modern Physics*, D. Reidel Publishing Co., Dordrecht, Holland (1976).

[12] The material quantum dialog-game was investigated in refs. 13, 14, 15.

[13] E.W. Stachow, Dissertation, Köln (1975).

[14] E.W. Stachow, *J. Philos. Logic* **5** 237 (1976).

[15] P. Mittelstaedt and E.W. Stachow, *J. Philos. Logic* **7** 181 (1978).

CHAPTER 4

THE CALCULUS OF EFFECTIVE QUANTUM LOGIC

In this chapter, the calculus of effective (intuitionistic) quantum logic will be established. Starting from the observation that there are some material propositions which can be defended within the material dialog-game D_m^* irrespective of the elementary propositions contained in it, the concept of a formally true proposition will be defined in Section 4.1 by the respective argument rules. Eliminating elementary propositions but not commensurabilities in D_m the semi-formal quantum dialog-game $D_f^{(m)}$ is formulated in Section 4.2. Following this, the material availability propositions are also eliminated, thus achieving the formal quantum dialog-game D_f (Section 4.3). The totality of propositions which can be defended in this formal dialog-game is comprehended in the calculus Q_{eff} of effective quantum logic, which is established in Section 4.4. This calculus Q_{eff} can be shown to be consistent and complete with respect to those propositions which can be defended in a formal dialog.

4.1 FORMALLY TRUE PROPOSITIONS

In the framework of the material dialog-games D_m and D_m^* (i.e. by means of the argument rules $A_m(1)$–$A_m(5)$) material dialogs can be performed in order to prove or disprove compound propositions. The question whether a certain compound proposition A can successfully be defended in a material dialog will, in general, depend on the truth or falsity of the material propositions contained in A. There are, however, a few compound propositions which can be defended in a material dialog irrespective of the material propositions contained in them. In order to illustrate this important fact, we shall consider here two examples, one of them belonging to the game D_m, the other one to D_m^*.

As the first example, we choose the proposition $A \equiv (a \wedge b) \rightarrow (a \rightarrow b)$, where a and b are arbitrary elementary propositions. The dialog for A within the game D_m then reads, for instance as (4.1). Here we have assumed that the opponent can prove the propositions

	O		P	
(0)	[ ]		$(a \wedge b) \to (a \to b)$	
(1)	$a \wedge b$	(0)	[ ]	
(2)	a	⟨1⟩	1?	(1)
(3)	a!	⟨2⟩	a?	(2)
(4)	b	⟨1⟩	2?	(1)
(5)	b!	⟨4⟩	b?	(4)
(6)	$k(a, b)$!	⟨1⟩	$k(a, b)$?	(1)
(7)	[ ]		$a \to b$	⟨0⟩
(8)	a	(7)	b	⟨7⟩
(9)	$k(a, b)$?	(7)	$k(a, b)$!	⟨7⟩
(10)	b?	(8)	b!	⟨8⟩

(4.1)

a, b and $k(a, b)$ in Rows $O(3)$, $O(5)$ and $O(6)$, respectively. This is the situation most disadvantageous for P. Nevertheless, P wins the dialog. He can prove the propositions $k(a, b)$ and b, which are necessary for a successful defence, in $P(9)$ and $P(10)$, since precisely these propositions have been shown to be valid previously by the opponent. Furthermore, for the proof of b in $P(10)$, availability problems do not occur. The proposition a, which is asserted in $O(8)$, does not influence the availability of proposition b, which has been proved in $O(5)$ – since the availability proposition $k(a, b)$ has also been proved in $O(6)$. The proponent can either refer to the result of Row $O(5)$ or he can repeat the proof of b, for which he gets the same result. Hence, P wins the dialog. If, on the other hand, O cannot prove any one of the propositions a, b or $k(a, b)$, he loses the dialog in Rows 3, 5 or 6, respectively.

The second example which we shall discuss is the proposition $A \equiv (a \wedge k(a, b)) \to (b \to a)$. Since this proposition contains explicitly the commensurability $k(a, b)$, the dialogic proof of A has to be performed within the extended material dialog-game D_m^*. The dialog of A then reads, for instance, as (4.2). Here we have assumed that the opponent can actually prove the material propositions a and $k(a, b)$ in Rows $O(3)$ and $O(5)$, respectively. This is again the situation which is most disadvantageous for P. Nevertheless, the proponent wins the dialog. First, he can prove $k(b, a)$ in $P(8)$, since according to the definition of the commensurability, $k(b, a)$ can be proved if and only

	O		P	
(0)	[ ]		$(a \wedge k(a, b)) \to (b \to a)$	
(1)	$a \wedge k(a, b)$	(0)	[ ]	
(2)	a	⟨1⟩	1?	(1)
(3)	$a!$	⟨2⟩	$a?$	(2)
(4)	$k(a, b)$	⟨1⟩	2?	(1)
(5)	$k(a, b)!$	⟨4⟩	$k(a, b)?$	(4)
(6)	[ ]		$b \to a$	⟨0⟩
(7)	b	(6)	a	⟨6⟩
(8)	$k(b, a)?$	(6)	$k(b, a)!$	⟨6⟩
(9)	$a?$	(7)	$a!$	⟨7⟩

(4.2)

if $k(a, b)$ can be proved – and $k(a, b)$ has been proved by the opponent in Row $O(5)$. Secondly, the proponent can prove a in $P(9)$ since the proposition a has already been shown to be valid by the opponent. On the other hand, the availability of a is not affected by the assertion of proposition b in Row $O(7)$, since the commensurability $k(b, a)$ has also been proved. Hence, the proponent wins this dialog. If, however, O cannot prove any one of the propositions a or $k(a, b)$ he loses the dialog as early as Rows 3 or 5, respectively.

These examples show that there are, in fact, compound propositions which can be defended in the material dialog-games D_m and D_m^* irrespective of the elementary propositions and of the commensurabilities contained in them. Propositions of this kind will be called *formally true.* In the following, formal procedures will be established which enable the construction of all formally true propositions.[1]

In principle, one could investigate the formal truth of a given proposition by considering all possibilities of a dialog with respect to the proof results of the *material* propositions. However, in general, it would be difficult to examine all such possibilities and to formalize this procedure in an appropriate manner. Therefore, we shall choose here a different way. We formulate the general possibilities of proving *formally true* propositions in a dialog, using considerations which have been applied already in the examples mentioned above. These possibilities will then be summarized in formal argument-rules which constitute the so-called *formal quantum dialogs.* A certain dialog-game D_x, which is defined by some argument-rules, determines a corresponding concept of truth. A proposition A is said to be x-true if P has a strategy of success in D_x for A. Here, we are going the opposite way. Starting from the concept of *formal truth*, we formulate the argument-rules A_f of propositions which are formally true (f-true). These argument-rules then determine a *formal quantum dialog-game* which will be denoted by D_f.

We have introduced, here, the material dialog-games D_m and D_m^*, which are related to the sets of propositions S and S^*, respectively. In the dialog-game D_m the material propositions $k(A, B)$ and $\bar{k}(A, B)$ are only contained implicitly as availability propositions whereas in D_m^* the commensurabilities k and $\bar{k}$ also appear as real propositions. Correspondingly, one should distinguish propositions which are *formally true* with respect to the dialog-games D_m and D_m^* and formal dialog-games D_f and D_f^*, respectively. In this chapter, we are mainly interested in establishing a calculus of effective (intuitionistic) quantum logic which can be compared formally with the calculus of ordinary effective logic. Since it turns out that this calculus of effective quantum logic is intimately related to D_f – and not to D_f^* – we restrict ourselves in this chapter to the investigation of the formal dialog-game D_f. The extended formal dialog-game D_f^* and the respective propositional calculus will be treated in Chapter 6.

The general validity of a formally true proposition $A \in S$ means that P wins the dialog independently of the evidence for the elementary propositions contained in A. Therefore, a proposition is *formally true* if P has a strategy of success even in the situation which is most unfavourable for him. This is the case if elementary propositions asserted by the opponent can always successfully be proved by him. Consequently, in the formal dialog-game we shall assume that O can, in fact, prove all elementary propositions proposed by him. On the other hand, a strategy of success can only exist if the proponent can never be committed by the opponent to prove a proposition explicitly – since the proposition in question could be false and in that case the proponent would lose the dialog. This is the case if P asserts only such propositions which he can take over from the opponent who can then not question them. Hence, we shall assume that P may only assert elementary propositions that have already been deployed by O in a dialog.

The proponent should, however, be allowed to refer to an elementary proposition which has been asserted previously by the opponent *only* if the respective proposition is *available*. As mentioned above, this means that the proof result obtained previously is still valid in the respective position of the dialog. The availability of a proposition a is given if, between the proof of a and its citing, no proposition incommensurable with a has been proved. In this way, the proof-procedures of all elementary propositions can, in fact, be eliminated in the dialog-game. There are, however, still material proof-procedures which have to be performed in the dialogs. At the present stage of our discussion, the availability of a proposition A which has been asserted by the opponent can only be demonstrated by proving all the material propositions $k(A, B_i)$ $(i = 1, \ldots n)$, where B_i are the propositions which have been stated between the proof of A and its citing.

We thus arrive at a dialog-game which is partly formal – since the proofs of elementary propositions have been eliminated – and partly material – since it still contains the availability propositions k and $\bar{k}$, the proofs of which have to be carried out outside the dialog. We shall denote this semi-formal quantum dialog-game here by $D_f^{(m)}$. Its argument rules will be formulated in Section 4.2. However, in the formal quantum dialog-game D_f the proofs of *all* material propositions (i.e. elementary propositions and commensurabilities) must not appear anymore. Consequently, in the next step the availability propositions k and $\bar{k}$, the proofs of which have to be performed outside the dialog, will also be eliminated. In this way, we achieve the formal quantum dialog-game D_f, which no longer contains the proof of any material proposition. The argument-rules of D_f will be formulated in Section 4.3.

4.2 FORMAL DIALOGS WITH MATERIAL COMMENSURABILITIES

The considerations of the last section can now be used in order to formulate *argument-rules* for the formal *dialog-game*. As just mentioned, we start with the semi-formal dialog-game $D_f^{(m)}$. Again, the argument-rules summarize the possibilities of proving some kind of propositions, namely the *formally true propositions*. The first argument-rule reads:

$A_f(1)$ (a): Elementary propositions are not attackable.

(b): O is allowed to state elementary propositions in every position of the dialog. P is allowed to state elementary propositions only if they have been asserted by O previously and if they are still available in the respective position of the dialog.

Similarly, in the formal dialog-game one has to take into account restrictions in the proponent's possibilities to attack a proposition, which again come from the possible incommensurabilities of quantum mechanical propositions. After having stated a proposition in a dialog, a participant is committed to defend this proposition against an attack only as long as it is still available. Therefore, we must restrict P's possibilities to attack previous propositions of O by the following argument rule:

$A_f(2)$: P is allowed to attack propositions of O only as long as they are available in the respective position of the dialog.

For the logical connectives, we use the same dialogic definitions in $D_f^{(m)}$ and in D_f as in the material quantum dialog-game D_m. However, we shall not consider the availability propositions as possibilities of attacks and defences, as has been done in $A_m(4)$, since the material propositions $k(A, B)$ and $\bar{k}(A, B)$ will be eliminated in the formal dialog-game. Instead, we go back to the original definition of the logical connectives, the dialogs of which are, of course, infinite. Therefore, we obtain the argument-rule:

$A_f(3)$:	connective		attacks	defences
(a):	conjunction	$A \wedge B$	1? 2?	A B
(b):	disjunction	$A \vee B$	?	A B
(c):	material implication	$A \rightarrow B$	A	B
(d):	negation	$\neg A$	A	

According to the argument-rules $A_f(1)$ and $A_f(2)$ a certain proposition could be infinitely available; i.e. it could be attacked or taken over an infinite number of times. However, the proponent must have a strategy of success within a finite dialog. Hence, we postulate that P establishes a bound which limits the number of times he may take an elementary proposition over and attack a proposition of O. This can be done by choosing a number n ($n = 1, 2, \ldots$) at the beginning of the dialog, or more precisely, after having asserted the initial argument. Since the class of formally provable propositions will, in general, depend on this number n, we consider at first the dialog-game $D_f^{m(n)}$, which is determined by a fixed n. Later, we take the union of all dialog-games $D_f^{m(n)}$ with respect to n, thus eliminating the conventional parameter n in our language.

Since P is assumed to possess a strategy of success he can defend against all arguments of O. Hence, it is sufficient to consider a dialog in which the opponent asserts his most advantageous argument. Furthermore, it can be shown that his chances of success cannot be improved by the possibility of more than one attack. Therefore, we restrict the possibilities of P and O by the following argument-rule:

$A_f(4^{(n)})$ (a): P is allowed to take over the same elementary proposition at most n times. P is allowed to attack the same proposition of O at most n times.

(b): O is allowed to attack propositions of P at most once. O has to decide if he is going to defend against an attack by P or if he is going to attack himself. In the case that he defends he no longer has the right to attack; in the case that he defends this defence is no longer possible.

Within a dialog scheme, this rule can technically be applied in the following way. Each proposition of O is labelled by an availability index i which indicates the number of times the respective proposition is available. After having been stated by the opponent, each proposition obtains the availability index n. In the further course of the dialog, this index is reduced by one if the respective proposition is taken over or attacked by P.

More important than this technical and purely conventional bound on the availability is the restriction of the availability of a proposition which comes from the possible incommensurability of quantum mechanical propositions.[7] Here, we have to take into account that a proposition A, once proved in the dialog, will be available in a later position of the dialog only if A is commensurable with all propositions B_i which have been stated between the proof of A and its citing. Obviously, it will be sufficient to consider O's propositions since only the opponent is allowed to assert elementary propositions which can then be taken over from O by P. Therefore, if the opponent has stated a new proposition A, the availability indices of all previous propositions C_i of O have to be reduced to zero, except when a proposition C_0 is commensurable with A. In this case, the availability index of C_0 remains unchanged. For the demonstration of the commensurability of C_0 and A one has to perform a proof of the material proposition $k(C_0, A)$.

In order to formulate an argument-rule which incorporates these restrictions of the commensurability in an appropriate way, one has to investigate all positions in a dialog in which O states a new proposition. However, it turns out that there are only three positions p_1, p_2, p_3 in which the availability of a previous proposition C is actually questioned by the assertion of a new proposition A.

p_1: O attacks a material implication $A \to B$ of P by A.

	O	P
(i)	C	
$\vdots$	$\vdots$	
(j)		$A \to B$
$(j+1)$	$A(j)$	

(4.3)

p_2: O attacks a negation $\neg A$ of P by A.

	O	P
$\vdots$	$\vdots$	
(i)	C	
$\vdots$	$\vdots$	
(j)		$\neg A$
$(j+1)$	$A(j)$	

(4.4)

p_3: O defends a disjunction $A \vee B$ against an attack of P.

	O	P
$\vdots$	$\vdots$	$\vdots$
(i)	C	.
$\vdots$	$\vdots$	$\vdots$
(k)	$A \vee B$	.
$\vdots$	$\vdots$	$\vdots$
(j)		$?(k)$
$(j+1)$	$A\langle k\rangle$	

(4.5)

One could now demand that in these three positions p_1, p_2, p_3 the proponent P is committed to reduce the availability of C correctly in Row j. In the case that P does not reduce the availability of C

correctly, the opponent should be allowed to question the commensurability (i.e. to attack by $k(C, A)$?) instead of asserting a new proposition. Furthermore, it should be mentioned that P cannot improve his strategy of success in p_1, p_2, p_3 if he is allowed to reduce the availability index of C to zero. Thus we arrive at the following argument-rule:

$A_f^m(5)$ (a): In a position p_1, p_2, p_3 of a dialog, P is allowed to reduce the availability of all previous propositions of O to zero.

(b): In a position p_1, p_2, p_3 of a dialog, where O can state a new proposition A, O is allowed to attack by $k(C, A)$? instead of asserting A. Here, C is any previous proposition of O with non-zero availability. The defence against $k(C, A)$? consists in a demonstration of $k(C, A)$.

In this formulation, the argument rule A_f(5a–5b) is somewhat too restrictive for the proponent, and must be weakened if one wants to obtain *all* propositions which are formally true. The reason is as follows: the validity of a proposition A, which has been proved at a certain stage of the dialog, is in general not *completely* destroyed by the proof of a new proposition B, which is not commensurable with A, but only partially. This means that, usually, there is still some information left about A which can be used in the further course of the dialog. In fact, it could happen that even if A is not commensurable with B, there is some weakening A' of A (with $A \to A'$) which is commensurable with B, and which, for this reason, should be still available for the proponent. A proposition A which has this property will be said to be *partially commensurable*[8] with B. As an example of the concept of partial commensurability, we consider the propositions $A = a$ and $B = (a \vee b) \to c$, which are obviously not commensurable. However, the weakening $A' = a \vee b$ of A can be shown to be commensurable with B irrespective of the elementary propositions a, b and c. (A' is a weakening of A since $A \to A'$ is formally true.) Hence, proposition A is partially commensurable with B.

In principle, it would be desirable to find the strongest proposition A' with $A \to A'$. This is, however, in general, not possible and we will confine ourselves to quote at least one proposition of this kind. A formal procedure for preserving some commensurable part A' of a proposition A in the dialog-game is as follows: P is allowed to challenge O to assert the formally true proposition $A \to A'$, which will be denoted here by $(A \to A')$? In this way, the opponent is only committed to defend a proposition which is known to be formally true. That this additional possibility of defence is actually sufficient to treat partial commensurabilities in an appropriate manner can be illustrated by the dialog about the proposition $C = a \to \{((a \vee b) \to c) \to a \vee b\}$. Here, we have $A = a$, $A' = a \vee b$ and $B = (a \vee b) \to c$. Thus, the dialog reads as (4.6). According to the rules

	O		P	
(0)	[ ]		$a \to \{((a \vee b) \to c) \to (a \vee b)\}$	
(1)	a	(0)	$((a \vee b) \to c) \to (a \vee b)$	$\langle 0 \rangle$
(2*)	$(a \vee b) \to c$	(1)	$a \vee b$	$\langle 1 \rangle$
(3*)	?		[ ]	
(2)	$a \to (a \vee b)$		$(a \to (a \vee b))?$	
(3)	$a \vee b$	$\langle 2 \rangle$	a	(2)
(4)	$(a \vee b) \to c$	(1)	$a \vee b$	$\langle 1 \rangle$
(5)	?	(4)	[ ]	
(6)	a	$\langle 3 \rangle$	?	(3)
(7)	[ ]		a	$\langle 4 \rangle$

(4.6)

$A_f(5a, 5b)$ in Row 2* the opponent would attack by $(a \vee b) \to c$. Since this proposition is not commensurable with a, the availability of a must be reduced to zero. Hence, the proponent is not able to defend his defence $a \vee b$ and loses the dialog in Row 3*. However, if the proponent makes use of the new possibility and challenges the proponent to assert the formally true proposition $a \to (a \vee b)$ in Row 2, he has a strategy of success. In fact, since $a \to (a \vee b)$ can be shown to be commensurable with a, the availability of a is not reduced and P can take over this proposition in $P(3)$ and attack $a \to (a \vee b)$. Furthermore, since O's defence $a \vee b$ in $O(3)$ is commensurable with O's attack $a \vee b \to c$ in $O(4)$, the availability of $a \vee b$ has not to be reduced to zero. Hence, P can take over the proposition $a \vee b$ and defend it successfully in Rows 5, 6 and 7. In this way, he wins the dialog.

Thus, we find that the allowance for P to challenge a formally true proposition in some situations improves his possibilities of success. Since, on the other hand, the assertion of a formally true proposition should not influence the formal truth of another proposition, we will incorporate these new possibilities of defence into the argument-rules of D_f. In order to formulate this additional argument-rule, we first generalize the above considerations to the case of several propositions $A_1, A_2 \ldots$ the availability of which is questioned in a dialog. Here, the new defence consists of the challenge $((A_1 \wedge \ldots \wedge A_n) \to B)$?. Secondly, the subdialog which starts with this challenge and ends with O's defence B (Rows 2 and 3 in the example) will be shortened by O's proposition B, since the propositions $(A_1 \wedge \ldots \wedge A_n) \to B$ and A_j are not relevant in the further course of the dialog. Finally, the opponent should have the opportunity to examine the formal truth of the propositions $(A_1 \wedge \ldots \wedge A_n) \to B$. Hence, instead of asserting B, the opponent should be allowed to ask for a proof of this pretended formally true proposition. Summarizing these aspects, we arrive at the following argument-rule, which must be established in addition to $A_f(5a)$ and $A_f(5b)$:

$A_f(5)$ (c): In a position p_1, p_2, p_3 of a dialog, P is allowed to challenge a proposition B. Thereupon, O may either assert the proposition B and continue the dialog in Row $(j+1)$, or he may ask: $((A_1 \wedge \ldots \wedge A_n) \to B)$? $A_1 \wedge \ldots \wedge A_n$ is the conjunction of all previous propositions of O with non-zero availabilities. P has to defend against this attack by asserting the new initial argument $(A_1 \wedge \ldots \wedge A_n) \to B$.

Finally, we exclude infinite strategies, which are still possible according to $A_f(5c)$, by the additional rule:

$A_f(6^{(n)})$: P is allowed to make use of $A_f(5c)$ at most n-times.

By means of the argument-rules $A_f(1)$, $A_f(2)$, $A_f(3)$, $A_f(4^{(n)})$, $A_f^m(5)$, $A_f(6^{(n)})$ the semi-formal dialog-game $D_f^{(m)}$ is established. Whereas the proofs of elementary propositions have been completely eliminated in this dialog-game, proofs of availability propositions must still be performed according to rule $A_f^{(m)}(5b)$.

4.3 THE FORMAL DIALOG-GAME

In the material dialog-game D_m, there are two kinds of material propositions, the elementary propositions and the commensurabilities k and $\bar{k}$. In the formal quantum dialog-game D_f the proofs of all material propositions must be eliminated. The proofs of elementary propositions have been eliminated in the last section by establishing the argument-rules of the semi-formal quantum dialog game $D_f^{(m)}$. However, commensurability propositions are still contained in the semi-formal quantum dialogs. Consequently, in the next step commensurability propositions $k(A, B)$, which depend on the elementary propositions of which A and B consist, must also be eliminated.

There are, however, certain commensurabilities $k(A, B)$ which can be demonstrated irrespective of the elementary propositions contained in A and B. It is obvious that these '*formal commensurabilities*' must not be eliminated but rather incorporated into the rules of the formal dialog-game. On the other hand, *formal incommensurabilities* do not exist. In order to demonstrate a formal commensurability $k(A, B)$, one has to recall the definition of $k(A, B)$ mentioned above (Definition (3.14): Material dialogs about A and B are carried out in turn; if it is certain that all material dialogs about A and B have always the same result, the formal commensurability is proved.

As an example of this proof procedure, we consider the formal commensurability $k(A, A \rightarrow B)$. If P loses the dialog about A, then O must lose it within the dialog about $A \rightarrow B$, too. Since only A has been tested, a new dialog about A will give the same result. If P wins the dialog about A, we have to distinguish several cases with respect to the dialog about $A \rightarrow B$. If P cannot prove $k(A, B)$, he loses the dialog. However, the proof attempt for $k(A, B)$ will reproduce the result of the A-dialog. If P can prove $k(A, B)$ then it does not matter whether he wins the subsequent dialog about B or not. Since $k(A, B)$ has been proved, the preceding result will be obtained in a new A-dialog. Thus, we find that A and $A \rightarrow B$ are, in fact, formally commensurable. This proof-procedure is further illustrated by the flux-diagram in Fig. 4.1.

In a similar way, one finds that the commensurabilities $k(A, A)$, $k(A, k(A, B))$ and $k(A, k(B, A))$ are formally true.[10] In addition to these formal commensurabilities, some relations between commensurabilities can be shown to be true, irrespective of the proof results of the elementary propositions (e.g. $\vdash k(A, B) \rightsquigarrow \vdash k(B, A)$). These relations will be used here as rules of a calculus[12] K, the beginnings of which are given by the formal commensurabilities mentioned above. In this calculus, the rules will be denoted by the double-arrow '$\Rightarrow$'; several premises will be connected by a double comma ',,'. The calculus K of formal commensurabilities then reads:

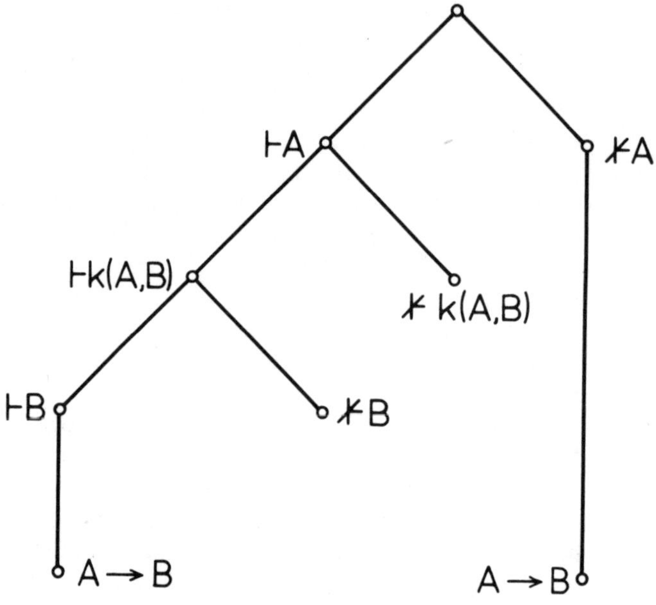

Fig. 4.1. Flux-diagram of the proof of $k(A, A \rightarrow B)$.

$K(1.1)$	$k(A, A)$
$K(1.2)$	$k(A, k(A, B))$
$K(1.3)$	$k(A, k(B, A))$
$K(1.4)$	$k(A, A \rightarrow B)$
$K(2)$	$k(A, B) \Rightarrow k(B, A)$
$K(3)$	$\vdash_{D(\gamma^m)} A \rightarrow B \Rightarrow k(A, B)$
$K(4)$	$k(A_1, B)$,, $k(A_2, B) \Rightarrow k(A_1 * A_2, B)$ with $* \in \{\wedge, \vee, \rightarrow\}$
$K(5)$	$k(A, B) \Rightarrow k(\neg A, B)$

This calculus may be considered as a formal tool which, starting from the beginnings $K(1.1)$–$K(1.4)$, allows for the construction of new formal commensurabilities. If a formal commensurability $k(A, B)$ can be derived in K, we write $\vdash_K k(A, B)$. It is not stated here that this calculus K is complete with respect to the method of proof for formal commensurabilities mentioned above. However, for the following considerations the calculus K will turn out to be sufficiently large.

Formal commensurabilities have to be proved by a derivation in the calculus K (i.e. by a proof procedure outside the dialog). Therefore, within the framework of the semi-formal dialog-game, formal commensurabilities must still be considered as material propositions. However, it can be shown that formal commensurabilities are in some sense equivalent to dialog-definite compound propositions. More precisely, the proof of a material (formal) commensurability proposition can be eliminated in any dialog. In fact, we have the following theorem:

(4.7) THEOREM: (a) If the formal commensurability $k(A, B)$ can be proved (i.e., $\vdash_K k(A, B)$) then there exists a strategy of success for the dialog about the proposition $A \to (B \to A)$.

(b) If there exists a strategy of success for the dialog about the proposition $A \to (B \to A)$, the obligation to prove $k(A, B)$ in a dialog can be circumvented.

Proof: (a) If $\vdash_K k(A, B)$ can be proved, then P has a strategy of success for $A \to (B \to A)$. In the dialog (4.8) P defends in $P(1)$ by

	O		P	
(0)	[ ]		$A \to (B \to A)$	
(1)	A	(0)	$B \to A$	$\langle 0 \rangle$
(2)	B	(1)	A	$\langle 1 \rangle$

(4.8)

asserting $B \to A$. Since $\vdash_K k(A, B)$ has been presupposed, O cannot attack by asking $k(A, B)$? Furthermore, since the availability of A in $O(1)$ is not changed by O's attack in $O(2)$, P can defend in $P(2)$ by taking over A from $O(1)$.

(b) The proof of the commensurability $k(A, B)$ may be challenged in the positions p_1, p_2, p_3, which are given by the dialogs (4.3), (4.4), (4.5), respectively. If, in the dialog (4.9), O has asserted A in Row m, and P asserts $B \to A$ in Row n, then O is allowed to attack by asking $k(A, B)$? according to A_f^m(5b). P can now circumvent this commensurability attack. Since, by assumption, P has a strategy of success for $A \to (B \to A)$, he makes use of A_f(5c) and – instead of defending by $k(A, B)!$ – he challenges the proposition $B \to A$. Thereupon, O may either assert $B \to A$ or he may ask: $A \to (B \to A)$? If O asks:

	O		P		
⋮	⋮		⋮		
(m)	A				
⋮	⋮		⋮		
(n)			$B \to A$		(4.9)
$(n+1)$	$k(A, B)$?		[ ]		
$(n+2^*)$	[ ]		$(B \to A)$?		
$(n+3^*)$	$A \to (B \to A)$?		$A \to (B \to A)$!		
$(n+2)$	$B \to A$!		$(B \to A)$?		
$(n+3)$	B	(n)	[ ]		
$(n+4)$	A	$\langle n+2 \rangle$	B	$(n+2)$	

$A \to (B \to A)$, P can defend by assumption (Rows $n+2^*$, $n+3^*$). In case O asserts $B \to A$, P is allowed to reduce the availability of A to zero. However, if O continues by asserting B as an attack against $P(n)$, P attacks by B. If O defends by A, P can take over this proposition and thus successfully defend the dialog.

Theorem (4.7) can be generalized to several propositions $A_1, \ldots A_n$, with non-zero availabilities which have been stated previously by the opponent. We thus arrive at the following generalization of (4.7):

(4.10) THEOREM: (a) If the formal commensurabilities $k(A_1, B)$, $k(A_2, B), \ldots$ can be proved then there exists a strategy of success for the proposition $(A_1 \wedge \ldots \wedge A_n) \to (B \to (A_1 \wedge \ldots \wedge A_n))$.

(b) If there exists a strategy of success for the proposition $(A_1 \wedge \ldots \wedge A_n) \to (B \to (A_1 \wedge \ldots \wedge A_n))$ the obligations to prove the commensurabilities $k(A_1, B)$, $k(A_2, B), \ldots$ can be circumvented.

The proof of this theorem consists of an obvious generalisation of the proof of Theorem (4.7) and will therefore not be presented here (cf. ref. 9).

On the basis of the Theorems (4.7) and (4.9), the material proofs from commensurability propositions $k(A, B)$ can be eliminated in any dialog and replaced by strategies for the respective propositions $A \to (B \to A)$. For this reason, the proposition $A \to (B \to A)$ will also be called the 'commensurability-connective' and denoted by

$c\{k(A, B)\} \equiv A \to (B \to A)$.[14] This leads to the following reformulation of the semi-formal argument-rule $A_f^m(5b)$ by the formal rule:

$A_f(5)$ (*b*): In a position p_1, p_2, p_3 of a dialog, where O can state a new proposition A, O is allowed to attack by $k(C, A)$? instead of asserting A. Here, C is the conjunction of all previous propositions of O with non-zero availabilities. The defence against $k(C, A)$? consists in asserting the new initial argument $C \to (A \to C)$.

Finally, in the formal dialog-game we have to incorporate all formal commensurabilities which are deducible within the calculus K. This can again be achieved by means of the equivalence theorems (4.7) and (4.9). We replace the beginnings of K (i.e. commensurabilities $k(A, B)$) by the respective new initial arguments $c\{k(A, B)\} \equiv A \to (B \to A)$ and the rules of K by rules concerning initial arguments. We thus arrive at the additional formal argument-rule:

$A_f(7)$ (a): O is not allowed to attack the initial argument $A \to ((A \to B) \to A)$.

(b): P is allowed to replace the initial argument $A \to (B \to A)$ by the new initial argument $B \to (A \to B)$.

(c): P is allowed to replace the initial argument $(A_1 * A_2) \to (B \to (A_1 * A_2))$ by the new initial arguments $A_1 \to (B \to A_1)$ or $A_2 \to (B \to A_2)$, with $* \in \{\wedge, \vee, \to\}$ and where O has to choose between the two possibilities.

(d): P is allowed to replace the initial argument $\neg A \to (B \to \neg A)$ by the new initial argument $A \to (B \to A)$.

(e): P is allowed to state the same initial argument at most once.

The rule $A_f(7e)$ is added here in order to exclude infinite strategies. The connectives, which correspond to the beginnings $K(1.1)$, $K(1.2)$, $K(1.3)$, are not mentioned in the argument-rule $A_f(7)$ since these propositions can be proved within the formal dialog-game. In fact, according to the argument-rules $A_f(1)$–$A_f(7)$ for the commensurability connectives

$$C\{k(A, A)\} \equiv A \to (A \to A),$$
$$C\{k(A, k(A, B))\} \equiv A \to ((A \to (B \to A)) \to A),$$
$$C\{k(A, k(B, A))\} \equiv A \to ((B \to (A \to B)) \to A),$$

there exist formal strategies of success. Furthermore, the rule $K(3)$ is not incorporated into the rule $A_f(7)$ since, within the formal dialog-game, it can easily be proved that $\vdash_{D_f} A \to B$ implies $\vdash_{D_f} A \to (B \to A)$.

In the argument-rules $A_f(1)$–$A_f(7)$ only formal commensurabilities are taken into account and material commensurabilities are completely eliminated. Hence, by the rules $A_f(1)$, $A_f(2)$, $A_f(3)$, $A_f(4^{(n)})$, $A_f(5)$, $A_f(6^{(n)})$, $A_f(7)$ the n-dialog-game $D_f^{(n)}$ is established. In order to obtain all formally true propositions, irrespective of the number n, we take the union of all dialog-games $D_f^{(n)}$ with respect to n. This game will be called the *formal quantum dialog-game* D_f.

4.4 THE CALCULUS Q_{eff} OF EFFECTIVE QUANTUM LOGIC

Propositions which can be defended in the formal quantum dialog-game D_f are said to be *formally true*. For a formally true proposition A the proponent possesses a strategy of success within the dialog-game D_f. If the proponent has a strategy of success against a proposition A, this proposition will be called *formally false*. On account of the argument-rule $A_f(3d)$ this is the case if and only if P has a strategy of success for $\neg A$. According to Definition (3.8) and in analogy to Definition (3.20), we use the following denotation:

(4.11) DEFINITION: $\vdash_{D_f} A \rightleftharpoons A$ is *formally true*; i.e. P has a strategy of success in D_f for A.

$\vdash_{D_f} \neg A \rightleftharpoons A$ is *formally false*; i.e. P has a strategy of success in D_f for $\neg A$.

Having now formulated the argument rules $A_f(1)$–$A_f(7)$ of the formal quantum dialog-game, we are going to establish a propositional calculus with the aid of which all those propositions can be derived which can be successfully defended in the formal quantum dialog game D_f. This calculus will be called the *calculus of effective quantum logic* Q_{eff}.[15] The formulae of Q_{eff} consist first of *beginnings* (i.e. assertions of the kind '$\vdash_{D_f} A$ is valid') and secondly of *rules* (i.e. implications of the kind $\vdash_{D_f} A \curvearrowright \vdash_{D_f} B$), which state that if a strategy of success for A is given then there exists a strategy of success for B. Within the calculus Q_{eff} we use the double arrow '$\Rightarrow$' for the designation of the rules (e.g. $\vdash_{D_f} A \Rightarrow \vdash_{D_f} B$). For assertions of the kind $\vdash_{D_f} A$ we sometimes use the variables $\alpha, \beta, \ldots$. If a rule contains several premises $\alpha, \beta, ., \ldots$ this will be denoted by the double comma ',,' (e.g. $\alpha \,,, \beta \Rightarrow \gamma$).

Beginnings $\vdash_{D_f} A$ must be established by a successful defence of A in a formal quantum dialog. A rule $\vdash_{D_f} A \Rightarrow \vdash_{D_f} B$ states that if proposition A can be defended in a formal quantum dialog then proposition B can also be justified dialogically. The dialogic proof of a rule $\vdash_{D_f} A \Rightarrow \vdash_{D_f} B$ will, therefore, be carried out in the following way: The proposition A will be presupposed by the opponent as a *hypothesis* before the dialog. The proponent has then to defend the proposition B in a formal dialog whereby he may refer to the hypothesis A which has been accepted by the opponent. It is obvious that the ability to refer to the hypothesis A is not restricted in any way. The restrictions of the availability which come from the general incommensurability of quantum mechanical propositions (cf. argument-rule $A_f(5)$) are not relevant for hypotheses.

In the schematic representation of formal dialogs the hypothesis will be treated in the following way: Propositions which are stated as hypotheses are placed before the real dialog in the new rows $O(-1)$ to $O(-m)$. In the formal (n)-dialog-game, hypotheses have the availability index n like new initial propositions. However, in contrast to initial propositions which are stated in Rows $1, 2, 3, \ldots$ the availability index of a hypothesis C is not reduced to zero due to the assertion of a proposition A incommensurable with C. Thus the proponent may refer to the hypothesis C in any position of the dialog irrespective of the commensurability of C with other propositions asserted by the opponent.

For a formulation of the calculus Q_{eff} which is most convenient for a comparison with a propositional lattice, it is useful to extend our formal language by introducing some special propositions and a 2-place relation. First, we augment the set S of propositions so far considered by the addition of the two propositions V (truth) and Λ (falsity). The use of both of these propositions within the framework of the dialogic method shall be established in such a way that V cannot be questioned by either participant of the dialog and that whoever maintains Λ shall have lost the dialog.

From this definition, it follows that the propositions $A \rightarrow \mathsf{V}$ and $\Lambda \rightarrow A$ can be defended in a dialog for all propositions A; i.e. these propositions are formally true. Indeed, in the dialogs (4.12), P has a strategy of success. The first dialog is lost by O since he attacks by asserting Λ. The second dialog is lost by O since he cannot attack V. Hence, we have proved $\vdash_{D_f} \Lambda \rightarrow A$ and $\vdash_{D_f} A \rightarrow \mathsf{V}$.

	O	P
(0)	[ ]	$\bigwedge \to A$
(1)	$\bigwedge$ (0)	

	O	P
(0)	[ ]	$A \to \bigvee$
(1)	A (0)	$\bigvee \langle 0 \rangle$

(4.12)

Furthermore, for the formulation of the calculus of effective quantum logic, it is useful to define, in addition to the operations $\wedge$, $\vee$, $\to$, $\neg$ on the set S of propositions, a 2-place relation $R \subseteq S \times S$ by

$$A \leqslant B \rightleftharpoons \vdash_{D_f} A \to B. \tag{4.13}$$

This relation will be called 'implication' and must be distinguished from the operation $A \to B$, denoted here as 'material implication'. According to the definition, the relation $A \to B$ between the propositions A and B holds if and only if the proposition $A \to B$ can be defended in a formal quantum dialog. If for two propositions the implications $A \leqslant B$ and $B \leqslant A$ are valid, we write $A = B$; i.e.:

(4.14) DEFINITION: $A = B \rightleftharpoons A \leqslant B$ and $B \leqslant A$.

Using the relation '$\leqslant$' it follows from the two dialogs (4.12) or from $\vdash_{D_f} \bigwedge \to A$ and $\vdash_{D_f} A \to \bigvee$, respectively, that we have the general validity of the implications

$$A \leqslant \bigvee, \qquad \bigwedge \leqslant A. \tag{4.15}$$

In particular, it follows from these relations together with the commensurability-rule $K(3)$, that $\bigvee$ and $\bigwedge$ are commensurable with all propositions. Hence, using Theorem (4.7) we obtain the relations

$$A \leqslant \bigvee \to A, \qquad A \leqslant \bigwedge \to A. \tag{4.16}$$

Furthermore, according to the dialogic definition of $\bigvee$ we find that for an arbitrary proposition A the statement $\vdash_{D_f} A$ is equivalent to $\vdash_{D_f} \bigvee \to A$. On the other hand, $\vdash_{D_f} \bigvee \to A$ is equivalent to the relation $\bigvee \leqslant A$ according to (4.13). Therefore, a proposition A can be defended in a formal quantum-dialog if and only if the relation $\bigvee \leqslant A$ holds; i.e.:

$$\vdash_{D_f} A \curvearrowleft\!\!\curvearrowright \bigvee \leqslant A. \tag{4.17}$$

From the dialogic definition of $\bigwedge$, we find that for an arbitrary proposition A the statement $\vdash_{D_f} \neg A$ is equivalent to $\vdash_{D_f} A \to \bigwedge$. Furthermore, since $\vdash_{D_f} A \to \bigwedge$ can be replaced by $A \leqslant \bigwedge$ we obtain the result

$$\vdash_{D_f} \neg A \curvearrowright A \leqslant \Lambda. \tag{4.18}$$

Applying (4.17) to proposition $\neg A$, we finally obtain

$$V \leqslant \neg A \curvearrowright A \leqslant \Lambda, \tag{4.19}$$

which means that A implies falsity if and only if $\neg A$ is true.

Using these formal tools, the calculus of effective quantum logic Q_{eff} can now be formulated in a most convenient way. For the formulae of the calculus we use combinations of the symbols Λ, V, A, B, C, ... with $\wedge$, $\vee$, $\rightarrow$, $\neg$ and $\leqslant$ and the bracket symbol. With these formulae we establish the beginnings and the rules of the calculus Q_{eff}. The 'beginnings' consist of implications $A \leqslant B$, whereas the 'rules' are formulae which allow for the derivation of further implications. The proofs for beginnings and rules of Q_{eff} will be performed by means of formal quantum dialogs.

(4.20) THEOREM: $A \leqslant A$.

Proof: P can defend in $P(1)$ by taking over A from $O(1)$.

	O	P
(0)	[ ]	$A \rightarrow A$
(1)	A (0)	$A \langle 0 \rangle$

(4.21) THEOREM: $A \leqslant B$,, $B \leqslant C \Rightarrow A \leqslant C$.

Proof: The propositions $A \rightarrow B$ and $B \rightarrow C$ are presupposed as hypotheses in Rows $O(-2)$ and $O(-1)$. P has then a strategy of success for $A \rightarrow C$. In Rows $P(2)$ and $P(3)$ the proponent can attack the hypothesis and thus take over proposition C for his defence in $P(4)$.

	O	P
(−2)	$A \rightarrow B$	
(−1)	$B \rightarrow C$	
(0)	[ ]	$A \rightarrow C$
(1)	$A(0)$	[ ]
(2)	$B \langle -2 \rangle$	A (−2)
(3)	$C \langle -1 \rangle$	B (−1)
(4)	[ ]	$C \langle 0 \rangle$

The mutual commensurabilities of A, B and C are irrelevant for this strategy of success.

(4.22) THEOREM: $A \wedge B \leqslant A$, $A \wedge B \leqslant B$.

Proof: P can defend in $P(3)$ by taking over A from $O(2)$ – irrespective of the commensurability of A and B. The proof for $A \wedge B \leqslant B$ is analogous.

	O	P
(0)	[ ]	$A \wedge B \to A$
(1)	$A \wedge B$ (0)	[ ]
(2)	$A \langle 1 \rangle$	1? (1)
(3)	[ ]	$A \langle 0 \rangle$

(4.23) THEOREM: $C \leqslant A$,, $C \leqslant B \Rightarrow C \leqslant A \wedge B$.

Proof: In $P(3)$ the proponent attacks by asserting C, which has been taken over from $O(1)$. Thereby, commensurability problems do

	O	P
(−2)	$C \to A$	
(−1)	$C \to B$	
(0)	[ ]	$C \to A \wedge B$
(1)	C (0)	$A \wedge B \langle 0 \rangle$
(2)	1? (1)	[ ]
(3)	$A \langle -2 \rangle$	C (−2)
(4)	[ ]	$A \langle 1 \rangle$
(5)	2? (1)	[ ]
(6)	$B \langle 1 \rangle$	C (−1)
(7)	[ ]	$B \langle 1 \rangle$

not occur. For the second attack in $P(6)$, P has to make sure that C is still available after the assertion of A in $O(3)$. Here, however, the first premise $C \leqslant A$ implies $\vdash_K k(C, A)$ according to the rule $K(3)$.

(4.24) THEOREM: $A \leqslant A \vee B$, $B \leqslant A \vee B$.

Proof: P can defend in $P(2)$ by taking over A from $O(1)$, irrespective of the commensurability of A and B. The proof for $B \to A \vee B$ is analogous.

	O	P
(0)	[ ]	$A \to A \vee B$
(1)	A (0)	$A \vee B$ ⟨0⟩
(2)	? (1)	A ⟨1⟩

(4.25) THEOREM: $A \leqslant C$,, $B \leqslant C \Rightarrow A \vee B \leqslant C$.

Proof: Here we have to distinguish whether O defends in $O(2)$ by A or by B. P then attacks in $P(3)$ by A or B, respectively. In each of

	O		P	
(−2)	$A \to C$			
(−1)	$B \to C$			
(0)	[ ]		$A \vee B \to C$	
(1)	$A \vee B$ (0)		[ ]	
(2)	A ⟨1⟩	B⟨1⟩	? (1)	
(3)	C ⟨−2⟩	C ⟨−1⟩	A (−2)	B (−1)
(4)			C ⟨0⟩	C ⟨0⟩

these cases O defends by C, which can then be taken over by P for his defence in $P(4)$. Commensurability problems do not occur in this dialog.

(4.26) THEOREM: $A \wedge (A \to B) \leqslant B$ (modus ponens law).

Proof: Here, the formal commensurability $k(A, A \to B)$ is used in

	O	P
(0)	[ ]	$(A \wedge (A \to B)) \to B$
(1)	$A \wedge (A \to B)$ (0)	[ ]
(2)	A ⟨1⟩	1? (1)
(3)	$A \to B$ ⟨1⟩	2? (1)
(4)	B ⟨3⟩	A (3)
(5)	[ ]	B ⟨0⟩

$P(4)$, when P takes over proposition A from $O(2)$. In the subsequent step P defends by B, which has been taken over from $O(4)$.

(4.27) THEOREM: $A \wedge C \leqslant B \Rightarrow A \to C \leqslant A \to B$.

Proof: Here, the opponent presupposes the hypothesis $A \wedge C \to B$ in Row (−1). In $P(3)$ the modus ponens law (4.26) has been used,

	O	P
(−1)	$A \wedge C \to B$	
(0)	[ ]	$(A \to C) \to (A \to B)$
(1)	$A \to C$ (0)	$A \to B$ ⟨0⟩
(2)	A (1)	[ ]
(3)	[ ]	$A \wedge C$ (−1)
(4)	2? (3)	[ ]
(5)	C ⟨1⟩	A (1)
(6)	B ⟨−1⟩	C ⟨3⟩
(7)	[ ]	B ⟨1⟩

which, however, could also have been proved in a subdialog. It is essential for this proof that A and C need not be commensurable.

Remark: In case A and C are commensurable, we have $C \leqslant A \to C$ and the rule (4.27) can be strengthened to

$$A \wedge C \leqslant B \Rightarrow C \leqslant A \to B. \tag{4.27*}$$

However, this rule, which is well-known from ordinary effective logic, cannot be demonstrated in the formal quantum dialog-game. It is, indeed, the decisive difference between ordinary logic and quantum logic. This difference will be investigated from an algebraic point of view in the next chapter.

(4.28) THEOREM: $A \leqslant B \to A \Rightarrow B \leqslant A \to B$.

Proof: According to the argument rule A_f(7b) P may replace the initial argument $B \to (A \to B)$ by $A \to (B \to A)$.

	O	P
(−1)	$A \to (B \to A)$	
(0)	[ ]	$B \to (A \to B)$
(1)	?	$A \to (B \to A)$

(4.29) THEOREM: $A_1 \leqslant B \to A_1$,, $A_2 \leqslant B \to A_2 \Rightarrow (A_1 * A_2) \leqslant B \to A_1 * A_2$.

Proof: According to the argument rule A_f(7c), P may replace the initial argument in $P(0)$ by one of the arguments in $O(-2)$ or $O(-1)$, respectively.

	O	P
(−2)	$A_1 \to (B \to A_1)$	
(−1)	$A_2 \to (B \to A_2)$	
(0)	[ ]	$(A_1 * A_2) \to (B \to A_1 * A_2)$
(1)	?	$A_1 \to (B \to A_1)$
		$A_2 \to (B \to A_2)$

(4.30) THEOREM: $A \wedge \neg A \leqslant \Lambda$ (law of contradiction).

Proof: Using (4.19), we show that there is a strategy of success for $\neg(A \wedge \neg A)$. Since A and $\neg A$ are formally commensurable, in Row $P(4)$

	O	P
(0)	[ ]	$\neg(A \wedge \neg A)$
(1)	$A \wedge \neg A$ (0)	[ ]
(2)	$A \langle 1 \rangle$	1? (1)
(3)	$\neg A \langle 1 \rangle$	2? (1)
(4)	[ ]	A (3)

the proponent can take over A from $O(2)$ and attack O's proposition $\neg A$ in $O(3)$. Thus, he wins the dialog.

(4.31) THEOREM: $A \wedge C \leqslant \Lambda \Rightarrow A \to C \leqslant \neg A$.

Proof: According to (4.19), we use $\neg(A \wedge C)$ as hypothesis. In $P(3)$, the proponent asserts $A \wedge C$ and thereby attacks the hypothesis. O has no possibility of attack against A and C, since both of these propositions are available for P. Hence, O loses the dialog.

Remark: In the case that A and C are commensurable, we have $C \leqslant A \to C$ and the rule (4.31) can be strengthened to

$$A \wedge C \leqslant \Lambda \Rightarrow C \leqslant \neg A \tag{4.31*}$$

	O	P
(−1)	$\neg(A \wedge C)$	
(0)	[ ]	$(A \to C) \to \neg A$
(1)	$A \to C$ (0)	$\neg A \langle 0 \rangle$
(2)	A (1)	[ ]
(3)	[ ]	$A \wedge C$ (−1)
(4)	2? (3)	A (1)
(5)	$C \langle 1 \rangle$	$C \langle 3 \rangle$

Just as the rule (4.27), this rule is also known from ordinary effective logic and cannot be demonstrated in the formal quantum dialog-game.

(4.32) THEOREM: $A \leq B \to A \Rightarrow \neg A \leq B \to \neg A$.

Proof: According to the argument rule A_f(7d) the proponent may replace the initial argument $\neg A \to (B \to \neg A)$ by the new argument

	O	P
(−1)	$A \to (B \to A)$	
(0)	[ ]	$\neg A \to (B \to \neg A)$
(1)	?	$A \to (B \to A)$

$A \to (B \to A)$. Since this argument is stated as hypothesis, P can successfully defend it.

Summarizing the beginnings and rules which have been demonstrated in Theorems (4.20)–(4.32) – apart from (4.27*) and (4.31*) – we will now present the calculus Q_{eff} of effective quantum logic. The calculus reads:

Q_{eff} (1.1) $A \leq A$,

Q_{eff} (1.2) $A \leq B$,, $B \leq C \Rightarrow A \leq C$.

Q_{eff} (2.1) $A \wedge B \leq A$,

Q_{eff} (2.2) $A \wedge B \leq B$,

Q_{eff} (2.3) $C \leq A$,, $C \leq B \Rightarrow C \leq A \wedge B$.

Q_{eff} (3.1) $A \leq A \vee B$,

Q_{eff} (3.2) $B \leq A \vee B$,

Q_{eff} (3.3) $A \leq C$,, $B \leq C \Rightarrow A \vee B \leq C$.

Q_{eff} (4.1) $A \wedge (A \to B) \leq B$,

Q_{eff} (4.2) $A \wedge C \leq B \Rightarrow A \to C \leq A \to B$,

Q_{eff} (4.3) $A \leq B \to A \Rightarrow B \leq A \to B$,

Q_{eff} (4.4) $B \leq A \to B$,, $C \leq A \to C \Rightarrow B * C \leq A \to (B * C)$, $* \in \{\wedge, \vee, \to\}$.

Q_{eff} (5.0) $\bigwedge \leq A$,

Q_{eff} (5.1) $A \wedge \neg A \leq \bigwedge$,

Q_{eff} (5.2) $A \wedge C \leq \bigwedge \Rightarrow A \to C \leq \neg A$,

Q_{eff} (5.3) $A \leq B \to A \Rightarrow \neg A \leq B \to \neg A$.

In this calculus, the formulas Q_{eff} (1.1, 2.1, 2.2, 3.1, 3.2, 4.1, 5.0, 5.1) are beginnings and Q_{eff} (1.2, 2.3, 3.3, 4.2, 4.3, 4.4, 5.2, 5.3) are rules. These rules, which are used here for the formulation of the calculus, will be called *constitutive rules*. It is obvious that there are further rules which can be demonstrated within the framework of the formal quantum dialog-game D_f, and which possibly allow for the derivation of some implications which cannot be derived merely from the beginnings and the constitutive rules of Q_{eff}. These rules will be said to be *dialogically provable*. However, for the following discussions it is not required to consider the totality of dialogically provable rules, since the constitutive rules are sufficient in a certain sense.

The calculus of effective quantum logic Q_{eff} must be considered as a formal means which, starting from the beginnings, allows for the derivation of further formally true implications. If an implication $A \leq B$ can be derived within the calculus Q_{eff} we write $\vdash_{\overline{Q_{\text{eff}}}} A \leq B$. It follows from the dialogic proof procedure of the beginnings and the constitutive rules that every implication derivable from Q_{eff} can be proved in a formal quantum dialog: that is, we have:

(4.33) THEOREM: (Consistency of Q_{eff}.)

$$\vdash_{\overline{Q_{\text{eff}}}} A \leq B \curvearrowright \vdash_{\overline{D_f}} A \rightarrow B.$$

The proof is given by the proofs of Theorems (4.20)–(4.32). On account of Theorem (4.33), the calculus Q_{eff} will be called *consistent* with regard to the class of quantum dialogically provable implications. Furthermore, the calculus Q_{eff} is also *complete* with regard to this class; i.e. every quantum-dialogically provable implication can be derived from Q_{eff}. Therefore, we have

(4.34) THEOREM: (Completeness of Q_{eff}.)

$$\vdash_{\overline{D_f}} A \rightarrow B \curvearrowright \vdash_{\overline{Q_{\text{eff}}}} A \leq B.$$

For the proof of this completeness theorem, one has first to enu merate by a recursive scheme all positions of a dialog for which there exists a strategy of success within the formal quantum dialog-game $D_f^{(n)}$ characterized by the argument-rules $A_f(1)$–$A_f(7)$. Following this, this recursive scheme can be reformulated by a tableau calculus T_n which allows for the derivation of so-called tableaux which correspond to positions of success in the formal quantum dialog-game.

The union of all calculi T_n with respect n ($n < \infty$) is called the tableau calculus T_{eff}. Finally, one shows that the calculus T_{eff} is consistent and complete with respect to the calculus Q_{eff} of effective quantum logic. The entire proof of the Completeness Theorem (4.34) has been performed by E.W. Stachow and can be found in refs. 6 and 9.

NOTES AND REFERENCES

[1] The concept of *formal truth* has been the subject of many investigations. Within the framework of the dialog-game and with respect to ordinary logic, this problem has been treated in refs. 2–4. With respect to propositions about quantum mechanical systems, the concept of formal truth is considered, in particular, in refs. 5 and 6.

[2] P. Lorenzen, *Metamathematik*, Bibliographisches Institut, Mannheim (1962).

[3] W. Kamlah and P. Lorenzen, *Logische Propädeutik*, Bibliographisches Institut, Mannheim, (1973).

[4] K. Lorenz, *Arch. f. Math. Logik und Grundlagenforsch.* **11**, (1968).

[5] P. Mittelstaedt, *Philosophical Problems of Modern Physics*, D. Reidel Publishing Co., Dordrecht, Holland (1976).

[6] E.W. Stachow, *J. Philos. Logic* **5**, 237 (1976).

[7] The quantum mechanical measuring process and its implications for the dialog-game are treated in ref. 5, Chapters III and VI.

[8] Partial commensurabilities are considered in ref. 6 and in more detail in ref. 9.

[9] E.W. Stachow, Dissertation, Köln, (1976).

[10] Detailed proofs for formal commensurabilities can be found in refs. 9 and 11.

[11] H.M. Denecke, Dissertation, Köln, (1976).

[12] The concept of 'calculus' has been investigated in great detail in ref. 13, §6.

[13] P. Lorenzen, *Formal logic*, D. Reidel Publishing Co., Dordrecht, Holland (1965).

[14] If a proposition $A \in S$ is denoted for some reason by two different symbols A and B, we write $A \equiv B$. The sign '$\equiv$' thus denotes the identity of two elements.

[15] For the concept of 'calculus', cf. ref. 13. The calculus Q_{eff} of effective quantum logic is considered in refs. 5 and 6.

CHAPTER 5

THE LATTICE OF EFFECTIVE QUANTUM LOGIC

In this chapter, we investigate the algebraic structure of the calculus Q_{eff} of effective quantum logic. Since Q_{eff} can be shown to possess the property of syntactic completeness (Section 5.1) the calculus can be completely replaced by a lattice. This lattice will be called quasi-implicative and denoted by L_{qi}. In Section 5.2, some important properties of the lattice L_{qi} will be mentioned. Furthermore, we investigate the relation between the commensurability and implicative sublattices and show that the lattice L_{qi} is a relaxation of the implicative lattice L_i (Section 5.3). Finally, it will be shown that L_{qi} is also a relaxation of the orthocomplemented quasimodular lattice from which it differs by only the 'tertium non datur' law.

5.1 THE QUASI-IMPLICATIVE LATTICE L_{qi}

5.1.1 *Deducible and Admissible Rules*

The calculus of effective quantum logic given by Q_{eff}(1.1–5.3) consists of *beginnings* and *constitutive rules*. It is obvious that there are further rules which can be deduced from the constitutive rules by purely logical inference. These rules are said to be *deducible*. Deducible rules of the form $\alpha \Rightarrow \beta$ – with α, β, γ as variables for implications – are then to be proved in the following way: After having added the implication α to the beginnings of Q_{eff}, the implication β must be derived within the calculus Q_{eff}. Therefore, for a deducible rule we use the notation $\alpha \vdash_{Q_{\text{eff}}} \beta$. Apart from deducible rules there exists another kind of rule which in Q_{eff} must be considered as valid in a certain sense and which will be called *admissible rules.*[1] A rule $\alpha \Rightarrow \beta$ is said to be admissible if after the addition of $\alpha \Rightarrow \beta$ to the constitutive rules of Q_{eff} no implication γ can be deduced which is not also deducible in Q_{eff} itself. For an admissible rule, we use the notation $\vdash_{Q_{\text{eff}}} \alpha \Rightarrow \beta$.

It follows from the (semantic) completeness of Q_{eff} (i.e. from Theorem (4.34)) that any quantum dialogically provable rule is admissible. Conversely, the Consistency Theorem (4.33) guarantees that every admissible rule $\alpha \Rightarrow \beta$ of Q_{eff} can be proved quantum dialogically. In this case, one has to show that under the hypothesis α the implication β can be proved quantum dialogically. However, on account of Theorems (4.33) and (4.34), it is no longer necessary to make any reference to the dialogic technique. Hence, in the following discussions we shall be concerned exclusively with the calculus Q_{eff} and its extension by deducible and admissible rules.

Within the framework of the calculus Q_{eff}, the admissibility of a rule $\alpha \Rightarrow \beta$ has to be proved in the following way: According to the definition of 'admissible', the rule $\alpha \Rightarrow \beta$ is admissible if after the extension of the calculus Q_{eff} by this rule, no implication γ can be derived which cannot also be derived in Q_{eff} itself. Hence, for the proof of the admissibility, one has to show that the implication β can be derived from all premises $\alpha^{(1)}, \alpha^{(2)}, \ldots$ which are sufficient for a derivation of α, and that, exclusively, by the constitutive rules of Q_{eff}. If we denote an admissible rule by $\alpha \overset{(A)}{\Rightarrow} \beta$ and a deducible one by $\alpha \overset{(D)}{\Rightarrow} \beta$, the proof procedure of an admissible rule can be expressed in the following way:

$\alpha \overset{(A)}{\Rightarrow} \beta$ if and only if: for all premises p_α with $p_\alpha \overset{(D)}{\Rightarrow} \alpha$ it follows that $p_\alpha \overset{(D)}{\Rightarrow} \beta$

It is an obvious consequence of this proof procedure that any deducible rule is also admissible; i.e. we have the result

$$\alpha \vdash_{Q_{\text{eff}}} \beta \quad \text{implies} \quad \vdash_{Q_{\text{eff}}} \alpha \Rightarrow \beta. \tag{5.1}$$

5.1.2 *The Syntactical Completeness of Q_{eff}*

The implication (5.1) is valid for any arbitrary calculus. In addition, the calculus of effective quantum logic given by Q_{eff}(1.1–5.3) has the special and important property that a rule which is admissible with respect to the constitutive rules can also be deduced in Q_{eff}; i.e., we have the result:

(5.2) THEOREM: (Syntactical Completeness of Q_{eff}) In the calculus Q_{eff} of effective quantum logic $\vdash_{Q_{\text{eff}}} \alpha \Rightarrow \beta$ implies $\alpha \vdash_{Q_{\text{eff}}} \beta$.

This property will be called *syntactical completeness*. It means that the beginnings and the constitutive rules of Q_{eff} are complete in the sense that starting from these rules all admissible rules can be deduced. Therefore, it is no longer necessary to consider special proof procedures for the admissibility.[2]

The fact that the syntactical completeness is a special property of Q_{eff} and not a general property of any calculus can easily be demonstrated by a counter-example. Consider, for example, the calculus $\tilde{Q}_{\text{eff}}$ which is given by $Q_{\text{eff}}(1.1\text{–}4.2)$ and the additional rule

$$A \leqslant B \Rightarrow \mathsf{V} \leqslant A \to B. \tag{5.3}$$

Since (5.3) is a deducible rule of Q_{eff}, the calculus $\tilde{Q}_{\text{eff}}$ is a subcalculus of Q_{eff}; i.e. the constitutive rules of $\tilde{Q}_{\text{eff}}$ are deducible in Q_{eff}. It can easily be shown that in the calculus $\tilde{Q}_{\text{eff}}$ the rule

$$A \leqslant B \to A \Rrightarrow \mathsf{V} \leqslant B \to A \tag{5.4}$$

is admissible. However, it is not deducible. If it were deducible in $\tilde{Q}_{\text{eff}}$ it would also be deducible in Q_{eff}. In Q_{eff}, however, it is easy to find a counter-example of (5.4): The implication $A \leqslant (A \vee B) \to A$ can be derived in Q_{eff} (cf. Theorem (5.10c)). Thus the premise of (5.4) is generally fulfilled. However, the conclusion $\mathsf{V} \leqslant (A \vee B) \to A$ is not valid for arbitrary propositions A and B. Hence (5.4) is not generally valid.

The example of the calculus $\tilde{Q}_{\text{eff}}$ shows that it is indeed necessary to demonstrate the syntactical completeness of Q_{eff}. The proof of this property has to be performed in the following way: Let us assume that $\alpha \Rrightarrow \beta$ is admissible, i.e. $\alpha \overset{(A)}{\Rrightarrow} \beta$. This means that for all premises $\alpha^{(n)}$ with $\alpha^{(n)} \overset{(D)}{\Rrightarrow} \alpha$ it follows that $\alpha^{(n)} \overset{(D)}{\Rrightarrow} \beta$. Then one has to show that from this assumption it follows that $\alpha \overset{(D)}{\Rrightarrow} \beta$. This proof procedure must now be carried out for all possible implications α. As a simple example, we consider the case where α has the form:

$$\alpha \equiv A \leqslant C \wedge D.$$

The possible premises are (according to $Q_{\text{eff}}(2.3)$) $A \leqslant C$ and $A \leqslant D$. Hence $\alpha \overset{(A)}{\Rrightarrow} \beta$ means: $A \leqslant C,, A \leqslant D \overset{(D)}{\Rrightarrow} \beta$. From $Q_{\text{eff}}(2.1, 2.2, 2.3)$ it follows that $A \leqslant C \wedge D \overset{(D)}{\Rrightarrow} A \leqslant C,, A \leqslant D$, and hence $\alpha \equiv A \leqslant C \wedge D \overset{(D)}{\Rrightarrow} \beta$, completing the proof.]

The complete proof of Theorem (5.2) has to take account of all possible implications α. Hence, it is rather tedious and – for this reason – will not be presented here. It can be found in the literature.[3]

5.1.3 *The Lattice of the Calculus* Q_{eff}

The property of syntactical completeness has the consequence that the calculus Q_{eff} can be completely replaced by a lattice. This lattice will then have the calculus Q_{eff} as a model. In order to transform the calculus Q_{eff} into a lattice, we use the following translation scheme:

Calculus	*Lattice*
proposition A	lattice element A
implication $A \leqslant B$	partial-ordering relation $A \leqslant B$
connectives $\wedge, \vee$	lattice operations $\wedge, \vee$
connectives $\rightarrow, \neg$	further lattice operations $\rightarrow, \neg$
rules $\alpha \Rrightarrow \beta$	lattice axioms $\alpha \curvearrowright \beta$

It follows from the rules $Q_{\text{eff}}(1.1\text{–}1.2)$ that the implication $A \leqslant B$ can be considered as a partial-ordering relation on the set S of propositions if the double arrow '$\Rrightarrow$' can be interpreted as logical inference. Under this assumption, the connectives $A \wedge B$ and $A \vee B$ are then operations on S which, according to $Q_{\text{eff}}(2.1\text{–}3.3)$, correspond to the infimum and supremum of A and B with respect to the partial-ordering relation. Hence, the rules $Q_{\text{eff}}(1.1\text{–}3.3)$ are axioms which constitute a lattice. The rules $Q_{\text{eff}}(4.1\text{–}5.3)$ then further specialize this lattice by postulating a zero element Λ and by defining the operations $A \rightarrow B$ and $\neg A$. The lattice which is obtained in this way will be called *quasi-implicative* lattice and will be denoted by L_{qi}. (The reason for this terminology will become clear in Section 5.3.)

It is obvious that the calculus Q_{eff} can be completely replaced by the lattice L_{qi} only if the double arrow '$\Rrightarrow$' can actually be considered as a logical inference; i.e., if the calculus Q_{eff} is *syntactically complete.* In fact, a lattice theoretical statement of the form $\alpha \curvearrowright \beta$, which can be deduced from the axioms of the lattice L_{qi}, will, in general, correspond to a deducible rule $\alpha \overset{(D)}{\Rrightarrow} \beta$ of the calculus Q_{eff}. Admissible rules $\alpha \overset{(A)}{\Rrightarrow} \beta$ of Q_{eff} are represented by deducible propositions $\alpha \curvearrowright \beta$ of the lattice L_{qi} only if Q_{eff} is syntactically complete, i.e. if an admissible rule is also deducible. Otherwise, the admissible rules of Q_{eff} would be lost in replacing the calculus by a lattice and thus the replacement would not be complete. Therefore, it is a consequence of Theorem (5.2) that the calculus Q_{eff} of effective quantum logic can actually be replaced by the lattice L_{qi}.

5.1.4 *The Axioms of the Quasi-implicative Lattice*

The quasi-implicative lattice L_{qi} is given by axioms which are formally equivalent to the rules $Q_{eff}(1.1–5.3)$ and which will be denoted here by $L_{qi}(1.1–5.3)$, respectively. The axioms of L_{qi} read:

The lattice L_{qi} is a set $S = \{A, B, \ldots\}$ of elements with a partial-ordering relation $\leq$ such that

$L_{qi}(1.1)$ $\quad A \leq A,$

$L_{qi}(1.2)$ $\quad A \leq B, B \leq C \curvearrowright A \leq C.$

The equivalence relation '=' can be defined by

$L_{q}(1.3)$ $\quad A \leq B, B \leq A \curvearrowleft\!\!\curvearrowright A = B.$

Furthermore, for any two elements $A, B \in L_{qi}$, and with respect to the relation $\leq$, there exists an element $A \wedge B$ (infimum) and an element $A \vee B$ (supremum) such that:

$L_{qi}(2.1)$ $\quad A \wedge B \leq A,$

$L_{qi}(2.2)$ $\quad A \wedge B \leq B,$

$L_{qi}(2.3)$ $\quad C \leq A, C \leq B \curvearrowright C \leq A \wedge B.$

$L_{qi}(3.1)$ $\quad A \leq A \vee B,$

$L_{qi}(3.2)$ $\quad B \leq A \vee B,$

$L_{qi}(3.3)$ $\quad A \leq C, B \leq C \curvearrowright A \vee B \leq C.$

In addition to these axioms, which constitute a lattice for any two elements $A, B \in L_{qi}$, there exists an element $A \to B$ called *quasi-implication* which satisfies the axioms:

$L_{qi}(4.1)$ $\quad A \wedge (A \to B) \leq B,$

$L_{qi}(4.2)$ $\quad A \wedge C \leq B \curvearrowright A \to C \leq A \to B,$

$L_{qi}(4.3)$ $\quad A \leq B \to A \curvearrowright B \leq A \to B,$

$L_{qi}(4.4)$ $\quad B \leq A \to B, C \leq A \to C \curvearrowright B * C \leq A \to (B * C),$
with $* \in \{\wedge, \vee, \to\}$.

Moreover, the lattice L_{qi} has a zero element Λ and for any element

$A \in L_{qi}$ there exists an element $\neg A$, which will be called *quasi-pseudocomplement* and which satisfies the axioms:

$L_{qi}(5.0)$ $\quad \Lambda \leqslant A$,

$L_{qi}(5.1)$ $\quad A \wedge \neg A \leqslant \Lambda$,

$L_{qi}(5.2)$ $\quad A \wedge C \leqslant \Lambda \curvearrowright A \rightarrow C \leqslant \neg A$,

$L_{qi}(5.3)$ $\quad A \leqslant B \rightarrow A \curvearrowright \neg A \leqslant B \rightarrow \neg A$.

Remark 1: The axioms $L_{qi}(4.1–4.4)$ are a weakening of the axioms of an implicative lattice. Hence L_{qi} is called 'quasi-implicative' lattice. Details can be found in Section 5.3.

Remark 2: The axioms $L_{qi}(5.0–5.3)$ are a weakening of the axioms which define the pseudocomplement.[4,5] Hence the element $\neg A$ is called "quasi-pseudocomplement".

5.2 PROPERTIES OF THE LATTICE L_{qi}

In this section, we mention some properties of the lattice L_{qi} which are of interest for the quantum logical interpretation of the lattice. On account of the Syntactical Completeness Theorem (5.2) all formal problems concerning the calculus Q_{eff} can be investigated on the lattice L_{qi}. Hence the results obtained in this section are equally valid for the calculus Q_{eff}.

According to $L_{qi}(5.0)$, in the lattice L_{qi} there exists a zero element. The existence of a unit element is not postulated by an axiom. However, it follows from the axioms of L_{qi} that in addition to the zero element, a unit element also exists in this lattice.

(5.5) THEOREM: In the lattice L_{qi} there exists a unit element V such that for any $A \in L_{qi}$ the relations

(a) $\quad A \leqslant V$

(b) $\quad V \leqslant A \rightarrow V$,

(c) $\quad V \leqslant A \rightarrow A$,

hold. The element V is given by $V = \neg \Lambda$.

Proof: It follows from $L_{qi}(5.0)$ that there exists a zero element Λ such that for any $A \in L_{qi}$ the relation $\Lambda \leqslant A$ holds. Hence, we obtain $\Lambda \leqslant A \rightarrow \Lambda$, and using $L_{qi}(4.3)$

$$A \leqslant \Lambda \rightarrow A. \tag{5.6}$$

From $\Lambda \leqslant A \to \Lambda$ it follows, on account of $L_{qi}(5.3)$, that

$$\neg\Lambda \leqslant A \to \neg\Lambda. \tag{5.7}$$

Furthermore, if we put $A = \Lambda$ in $L_{qi}(5.2)$ we obtain $\Lambda \to C \leqslant \neg C$, and by means of (5.6) it follows that

$$C \leqslant \neg\Lambda. \tag{5.8}$$

Hence, the element $\mathsf{V} = \neg\Lambda \in L_{qi}$ fulfills the relations (5.5a) and (5.5b).

To prove (5.5c) in $L_{qi}(4.2)$ we put $C = \mathsf{V}$ and $B = A$. Hence, it follows that $A \to \neg\Lambda \leqslant A \to A$, and using (5.7) we finally obtain $\neg\Lambda \leqslant A \to A$, completing the proof.⟧

The unit element V is a very useful formal tool in order to express lattice theoretical relations. The following propositions are particularly important:

(5.9) THEOREM: In the lattice L_{qi} the propositions

(a) $A \leqslant B \curvearrowleft\!\curvearrowright \mathsf{V} \leqslant A \to B$,

(b) $A \leqslant \Lambda \curvearrowleft\!\curvearrowright \mathsf{V} \leqslant \neg A$;

are deducible.

Proof: (a) $A \leqslant B$ implies $A \wedge \mathsf{V} \leqslant B$, and using $L_{qi}(4.2)$, $A \to \mathsf{V} \leqslant A \to B$. Applying (5.5b), it follows that $\mathsf{V} \leqslant A \to B$. From $\mathsf{V} \leqslant A \to B$ we obtain $A = A \wedge (A \to B) \leqslant B$ using $L_{qi}(4.1)$, and thus $A \leqslant B$.⟧

(b) $A \leqslant \Lambda$ implies $A \wedge A \leqslant \Lambda$, and using $L_{qi}(5.2)$, $A \to A \leqslant \neg A$. Applying (5.5c), it follows that $\mathsf{V} \leqslant \neg A$. From $\mathsf{V} \leqslant \neg A$ we obtain $A = A \wedge \neg A \leqslant \Lambda$, using $L_{qi}(5.1)$, and thus $A \leqslant \Lambda$.⟧

The formal commensurabilities, which can be derived within the calculus K, are incorporated into the formal dialog-game and in the calculus Q_{eff} of effective quantum logic. Hence, these formal commensurabilities are also contained in L_{qi}.

(5.10) THEOREM: (Formal Commensurabilities.) In the lattice L_{qi} the following relations hold:

(a) $A \leqslant A \to A$,

(b) $A \wedge B \leqslant A \to (A \wedge B)$,

(c) $A \vee B \leqslant A \to (A \vee B)$,

(d) $A \to B \leqslant A \to (A \to B)$,

(e) $\neg A \leqslant A \to \neg A$.

Proof: (a) From (5.5a) we obtain $A \leqslant \mathrm{V}$, from (5.5c) $\mathrm{V} \leqslant A \to A$ and hence $A \leqslant A \to A$.]]

(b) Using (5.9a), $A \wedge B \leqslant A$ (i.e. $L_{qi}(2.1)$) implies $A \leqslant \mathrm{V} \leqslant (A \wedge B) \to A$. Applying $L_{qi}(4.3)$ we obtain $A \wedge B \leqslant A \to (A \wedge B)$.]]

(c) Using (5.9a), $A \leqslant A \vee B$ (i.e. $L_{qi}(3.1)$) implies $A \vee B \leqslant \mathrm{V} \leqslant A \to (A \vee B)$.]]

(d) From $A \wedge (A \wedge B) \leqslant B$ it follows that $A \to (A \wedge B) \leqslant A \to B$, applying $L_{qi}(4.2)$. Using (5.10b) we obtain $A \wedge B \leqslant A \to B$, and again applying $L_{qi}(4.2)$, $A \to B \leqslant A \to (A \to B)$.]]

(e) From $A \leqslant A \to A$ (5.10a), we obtain $\neg A \leqslant A \to \neg A$ using $L_{qi}(5.3)$.]]

Many relations which are known to be valid in an implicative lattice can also be derived in L_{qi}. Here, we mention only a few of them:

(5.11) THEOREM: The following relations are valid in L_{qi}:

(a) $A \wedge B \leqslant A \to B$,

(b) $\neg A \leqslant A \to B$,

(c) $\neg A = A \to \Lambda$,

(d) $A \leqslant \neg\neg A$.

Remark: By (5.11c) the quasi-pseudocomplement is reduced to the quasi-implication. The inversion of (5.11d) is not deducible in L_{qi}, just as in an implicative lattice.

Proof: (a) Using $L_{qi}(4.2)$, from $A \wedge (A \wedge B) \leqslant B$ we obtain $A \to (A \wedge B) \leqslant A \to B$, and on account of (5.10b), $A \wedge B \leqslant A \to B$.]]

(b) From $A \wedge \neg A \leqslant B$ it follows that $A \to \neg A \leqslant A \to B$, using $L_{qi}(4.2)$, and on account of (5.10d) we obtain further $\neg A \leqslant A \to B$.]]

(c) If we put $B = \Lambda$ in (5.11b) we obtain $\neg A \leqslant A \to \Lambda$. Furthermore, from $A \wedge \Lambda \leqslant \Lambda$ it follows on account of $L_{qi}(5.3)$ that $A \to \Lambda \leqslant \neg\Lambda$.]]

(d) From (5.10e), we obtain by means of $L_{qi}(4.3)$ $A \leqslant \neg A \to A$, and from $\neg A \wedge A \leqslant \Lambda$, on account of $L_{qi}(5.2)$, $\neg A \to A \leqslant \neg\neg A$. Hence we obtain $A \leqslant \neg\neg A$.]]

In the formal dialog-game D_f the logical connectives $A \wedge B$, $A \vee B$, $A \to B$ and $\neg A$ are well-defined by the respective argument rules. Hence, one would expect that the lattice operations $\wedge$, $\vee$, $\to$, $\neg$ in L_{qi} – which is known to be complete with respect to D_f – are uniquely defined by the lattice axioms. This is actually the case.[6,7]

(5.12) THEOREM: In the lattice L_{qi} the operations $A \wedge B$, $A \vee B$, $\neg A$ and $A \to B$ are uniquely defined by the lattice axioms.

Proof: (a) *Uniqueness of* $A \wedge B$: Assume there exists a further operation $A * B$, which also fulfills $L_{qi}(2.1, 2.2, 2.3)$. Then it follows that

$$A * B \leqslant A, A * B \leqslant B \quad \text{implies} \quad A * B \leqslant A \wedge B,$$
$$A \wedge B \leqslant A, A \wedge B \leqslant B \quad \text{implies} \quad A \wedge B \leqslant A * B,$$

and we obtain $A \wedge B = A * B$.]]

(b) *Uniqueness of* $A \vee B$: Assume there exists a further operation $A * B$, which also fulfills $L_{qi}(3.1, 3.2, 3.3)$. Then it follows that

$$A \leqslant A * B, B \leqslant A * B \quad \text{implies} \quad A \vee B \leqslant A * B,$$
$$A \leqslant A \vee B, B \leqslant A \vee B \quad \text{implies} \quad A * B \leqslant A \vee B,$$

and we obtain $A \vee B = A * B$.]]

(c) *Uniqueness of* $\neg A$: Assume there exists a further operation A^* which also fulfills $L_{qi}(5.1, 5.2, 5.3)$. Then it follows that

$$A \wedge A^* \leqslant \Lambda \quad \text{implies} \quad A \rightarrow A^* \leqslant \neg A$$

and

$$A \leqslant A \rightarrow A \text{ (cf. (5.10a))} \quad \text{implies} \quad A^* \leqslant A \rightarrow A^*.$$

Hence, we obtain $A^* \leqslant \neg A$. The proof of the inverse relation $\neg A \leqslant A^*$ is analogous.]]

(d) *Uniqueness of* $A \rightarrow B$:

(5.13) LEMMA: $A \leqslant B \rightarrow A$ implies $A \leqslant f(A, B) \rightarrow A$, where $f(A, B)$ is an arbitrary operation generated by a finite number of combinations of $\wedge$, $\vee$, $\rightarrow$ and $\neg$.

Proof: From $A \leqslant B \rightarrow A$ and $A \leqslant A \rightarrow A$ (cf. (5.10a)) it follows by means of $L_{qi}(4.3)$ and $L_{qi}(4.4)$ for any finitely connected operation $f(A, B)$ that $A \leqslant f(A, B) \rightarrow A$.]]

Assume there exists a further operation $q(A, B)$ which also satisfies $L_{qi}(4.1, 4.2, 4.3, 4.4)$. Using $L_{qi}(4.2)$ it follows from $A \wedge B \leqslant A \rightarrow B$ (cf. (5.11a)) that $q(A, B) \leqslant q(A, A \rightarrow B)$ and from $A \wedge (A \rightarrow B) \leqslant B$ (cf. $L_{qi}(4.1)$) that $q(A, A \rightarrow B) \leqslant q(A, B)$. Hence, we obtain

$$q(A, B) = q(A, A \rightarrow B). \tag{5.14}$$

Furthermore, by means of Lemma (5.13), from the relation $A \leqslant (A \rightarrow B) \rightarrow A$ it follows that $A \leqslant q(A, A \rightarrow B) \rightarrow A$. Using (5.14) and

$L_{qi}(4.3)$ we obtain

$$q(A, B) \leqslant A \rightarrow q(A, B). \tag{5.15}$$

Applying $L_{qi}(4.2)$ to $A \wedge q(A, B) \leqslant B$ (cf. $L_{qi}(4.1)$) it follows that $A \rightarrow q(A, B) \leqslant A \rightarrow B$, from which, together with (5.15), we obtain finally $q(A, B) \leqslant A \rightarrow B$. The proof of the inverse relation $A \rightarrow B \leqslant q(A, B)$ is analogous.

In addition to the results mentioned in Theorem (5.11), the following lattice theoretical relations are of particular importance for the further investigations:

(5.16) THEOREM: The following relations are valid in L_{qi}:

(a) $A \leqslant B \curvearrowright A \leqslant B \rightarrow A$,

(b) $A \leqslant B \curvearrowright \neg B \leqslant \neg A$,

(c) $\neg A \vee \neg B \leqslant \neg(A \wedge B)$,

(d) $A \vee B \leqslant \neg(\neg A \wedge \neg B)$,

(e) $\neg(A \vee B) = \neg A \wedge \neg B$.

Remark: (α) (5.16a) corresponds to the rule $K(3)$ of the calculus of formal commensurabilities.

(β) The inverse relation to (5.16b) is not valid in L_{qi}.

(γ) (5.16c) and (5.16d) are weakenings of the de Morgan laws (2.2). The inverse relations of (5.16c) and (5.16d) are not valid in L_{qi}.

Proof: (a) $A \leqslant B$ implies $A \leqslant B \wedge A$, and on account of (5.11a), $A \leqslant B \rightarrow A$.]]

(b) $A \leqslant B$ implies $A \wedge \neg B \leqslant B \wedge \neg B \leqslant \Lambda$, and by means of $L_{qi}(5.2)$, it follows that $A \rightarrow \neg B \leqslant \neg A$. Furthermore, on account of (5.16a), $A \leqslant B$ implies $A \leqslant B \rightarrow A$ and $B \leqslant A \rightarrow B$, and using $L_{qi}(5.3)$, it follows that $\neg B \leqslant A \rightarrow \neg B$. Hence, we get $\neg B \leqslant \neg A$.]]

(c) From $A \wedge B \leqslant A$ and $A \wedge B \leqslant B$ it follows, on account of (5.16b) and $L_{qi}(3.3)$, that $\neg A \vee \neg B \leqslant \neg(A \wedge B)$.]]

(d) From $\neg A \wedge \neg B \leqslant \neg A$ and $\neg A \wedge \neg B \leqslant \neg B$ it follows, on account of (5.16b), $L_{qi}(3.3)$ and (5.11d), that $A \vee B \leqslant \neg\neg A \vee \neg\neg B \leqslant \neg(\neg A \wedge \neg B)$.]]

(e) From $A \leqslant A \vee B$, $B \leqslant A \vee B$ it follows, on account of (5.16b), (5.16d) and $L_{qi}(2.3)$, that $\neg(A \vee B) = \neg A \wedge \neg B$.

The lattice L_{qi} is not orthocomplemented and thus not quasimodular. However, a weakening of the quasimodular law, which uses the quasi-pseudocomplement instead of the orthocomplement, is still valid in L_{qi}.[6]

(5.17) THEOREM: (Weak Quasimodularity.) In the lattice L_{qi} the relation $B \leq A, C \leq \neg A \curvearrowright A \wedge (B \vee C) \leq B$ holds.

Proof: From (5.11a) and (5.11b) it follows that $\neg A \vee (A \wedge B) \leq A \rightarrow B$. On the other hand, $B \leq A$ implies $B \leq A \wedge B \leq \neg A \vee (A \wedge B)$ and $C \leq \neg A$ implies $C \leq C \wedge \neg A \leq \neg A \vee (A \wedge B)$. Hence it follows from the premises $B \leq A$ and $C \leq \neg A$ that $B \vee C \leq \neg A \vee (A \wedge B) \leq A \rightarrow B$, and thus that $A \wedge (B \vee C) \leq A \wedge (A \rightarrow B) \leq B$.]]

The lattice L_{qi} is not distributive. However, under two additional premises the distributive law can be proved also in L_{qi}.

(5.18) THEOREM: (Weak Distributivity.) In the lattice L_{qi} the relation $B \leq A \rightarrow B, C \leq A \rightarrow C \curvearrowright A \wedge (B \vee C) \leq (A \wedge B) \vee (A \wedge C)$ holds.

Remark: In the logical interpretation, the premises mean that B and C are both commensurable with A. The mutual commensurability of B and C is not required for the proof of the distributivity.

Proof: If we put $D = (A \wedge B) \vee (A \wedge C)$ it follows that

$$A \wedge B \leq D \curvearrowright A \wedge (A \wedge B) \leq D \curvearrowright A \rightarrow (A \wedge B) \leq A \rightarrow D$$
$$A \wedge C \leq D \curvearrowright A \wedge (A \wedge C) \leq D \curvearrowright A \rightarrow (A \wedge C) \leq A \rightarrow D$$

using $L_{qi}(4.2)$, and thus (on account of $L_{qi}(3.3)$)

$$(A \rightarrow (A \wedge B)) \vee (A \rightarrow (A \wedge C)) \leq A \rightarrow D.$$

Furthermore, again using $L_{qi}(4.2)$, we obtain $A \rightarrow B \leq A \rightarrow (A \wedge B)$ and $A \rightarrow C \leq A \rightarrow (A \wedge C)$, and hence

$$(A \rightarrow B) \vee (A \rightarrow C) \leq A \rightarrow D.$$

From the premises $B \leq A \rightarrow B$ and $C \leq A \rightarrow C$, it follows by means of this relation that

$$B \vee C \leq (A \rightarrow B) \vee (A \rightarrow C) \leq A \rightarrow D,$$

and thus

$$A \wedge (B \vee C) \leq A \wedge (A \rightarrow D) \leq D.]]$$

5.3 THE RELATION BETWEEN L_{qi} AND THE LATTICE L_i

In this section, we investigate the relation between the lattice L_{qi} of effective quantum logic and the implicative lattice L_i, which has as a model the Brower calculus of ordinary effective logic. The *implicative lattice with zero element* L_i which was already briefly mentioned in

Section 2.4 can be obtained if we replace the axioms $L_{qi}(4.1\text{–}4.4)$ and $L_{qi}(5.0\text{–}5.3)$ of the quasi-implicative lattice L_{qi} by the following postulates:[8]

In the lattice L_i, for any two elements $A, B \in L_i$, there exists an element $A \to B$ called *material implication*, which satisfies the axioms

$L_i(4.1)$ $\quad A \wedge (A \to B) \leq B,$

$L_i(4.2)$ $\quad A \wedge C \leq B \curvearrowright C \leq A \to B.$

Moreover, in the lattice L_i there is a zero element Λ and for any element $A \in L_i$ there exists an element $\neg A$, the *pseudocomplement* of A, which satisfies the axioms

$L_i(5.0)$ $\quad \Lambda \leq A,$

$L_i(5.1)$ $\quad A \wedge \neg A \leq \Lambda,$

$L_i(5.2)$ $\quad A \wedge C \leq \Lambda \curvearrowright C \leq \neg A.$

It is a consequence of these axioms that the material implication $A \to B$ as well as the pseudocomplement $\neg A$ are uniquely defined by the axioms $L_i(4.1\text{–}4.2)$ and $L_i(5.0\text{–}5.2)$, respectively. Implicative lattices have been investigated in full detail in the literature.[5] Here, we mention only that property of an implicative lattice which is most important for the following considerations: Any implicative lattice is distributive, i.e. for any three elements $A, B, C \in L_i$ the distributive law

$$A \wedge (B \vee C) = (A \wedge B) \vee (A \wedge C)$$

holds. Conversely, a finite distributive lattice or a completely distributive lattice (in which the distributive law holds even for an infinite series of elements) is implicative, i.e. the axioms $L_i(4.1\text{–}5.2)$ can be deduced. The proofs can be found in ref. 4. Since a quasi-implicative lattice L_{qi} is not distributive, but only weakly distributive, it follows that L_{qi} is – with respect to the distributivity – a weakening of the implicative lattice L_i. It can be seen from the following considerations that this relation between L_{qi} and L_i is not restricted to this special law.

According to the logical interpretation of the quasi-implicative lattice L_{qi}, the commensurability of two elements $A, B \in L_{qi}$ is given by the relation $A \leq B \to A$. In a similar way, as for a quasimodular lattice L_q (cf. Section 2.2), in the lattice L_{qi} the commensurability can also be defined as an abstract relation. This can be seen in the following way:

(5.20) DEFINITION: In the quasi-implicative lattice L_{qi} the 2-place relation $K \subseteq L_{qi} \times L_{qi}$ (commensurability) is defined by

$$(A, B) \in K \curvearrowleft\!\curvearrowright A \leq B \to A$$

and will be denoted by $A \sim B$.

It follows from the Axiom $L_{qi}(4.3)$ that this commensurability relation is *symmetric*, i.e.

$$A \sim B \curvearrowright B \sim A, \tag{5.21}$$

and from Theorem (5.16a), we find that the partial-ordering relation R which is given by '$\leq$' is below K, i.e.

$$R \leq K. \tag{5.22}$$

Furthermore, it is a consequence of Axioms $L_{qi}(4.4, 5.3)$ that the commensurability relation K is *closed* under the lattice operations $\wedge, \vee, \to, \neg$, i.e. we have

(a) $A \sim B, A \sim C \curvearrowright A \sim (B * C) \qquad * \in \{\wedge, \vee, \to\}$,

(b) $A \sim B \curvearrowright A \sim \neg B$.

Consequently, for three elements $A, B, C \in L_{qi}$ which are pairwise commensurable, i.e. for which the relations $A \sim B$, $A \sim C$, $B \sim C$ hold, it follows that the relations

$$\begin{aligned} &A \sim \neg B, A \sim \neg C, A \sim B * C, \\ &B \sim \neg A, B \sim \neg C, B \sim A * C, \\ &C \sim \neg A, C \sim \neg B, C \sim A * B, \end{aligned} \tag{5.23}$$

with $* \in \{\wedge, \vee, \to\}$, are also satisfied. This means that all elements of L_{qi} which are formed from A, B, C and arbitrary iterations of the operations $\wedge, \vee, \to, \neg$ are mutually commensurable.

Three elements $A, B, C \in L_{qi}$ which are pairwise commensurable thus generate a sublattice $L(A, B, C) \subseteq L_{qi}$ of elements which are mutually commensurable. On the other hand, for commensurable elements from $L_{qi}(4.2)$ and $L_{qi}(5.2)$, the stronger axioms $L_i(4.2)$ and $L_i(5.2)$, respectively, can be deduced, i.e.:

(5.24) LEMMA: For elements $A, B, C \in L_{qi}$ which are mutually commensurable, from $L_{qi}(4.2)$ and $L_{qi}(5.2)$, $L_i(4.2)$ and $L_i(5.2)$ follow, respectively.

Proof: (a) From $L_{qi}(4.2)$, namely $A \wedge C \leq B \curvearrowright A \to C \leq A \to B$ and $C \leq A \to C$, it follows that $A \wedge C \leq B \curvearrowright C \leq A \to B$.]]

(b) From $L_{qi}(5.2)$, namely $A \wedge C \leq \Lambda \curvearrowright A \to C \leq \neg A$ and $C \leq A \to C$, it follows that $A \wedge C \leq \Lambda \curvearrowright C \leq \neg A$.]]

Hence, we conclude that the sublattice $L(A, B, C) \subseteq L_{qi}$, which is generated by mutually commensurable elements of L_{qi}, is *implicative.* On the other hand, since from $L_i(4.2)$ the relation $B \leq A \to B$ follows, the elements $A, B \in L_i$ of an implicative lattice are always commensurable. Therefore, we arrive at the following result:

(5.25) THEOREM: (a) If three elements $A, B, C \in L_{qi}$ are pairwise commensurable, then the lattice which is generated by these elements is an implicative sublattice $L_i(A, B, C) \subseteq L_{qi}$ of the lattice L_{qi}.

(b) If L_i is an implicative sublattice of L_{qi}, i.e. $L_i \subseteq L_{qi}$ and if $A, B, C, \ldots$ are elements of L_i then these elements are pairwise commensurable.

These results concerning the commensurability relation K in a quasi-implicative lattice can be summarized in the following statement: In the lattice L_{qi} the relation $K \subseteq L_{qi} \times L_{qi}$ fulfills the following four conditions:

(K1) K is symmetric.

(K2) $R \subseteq K$.

(K3) If $\mathcal{S} \subseteq L_{qi}$ is a subset of elements with $\mathcal{S} \times \mathcal{S} \subseteq K$, then $\mathcal{S}$ generates an implicative sublattice $L_i(\mathcal{S}) \subseteq L_{qi}$.

(K4) If $L_i \subseteq L_{qi}$ is an implicative sublattice, the elements of any subset $\mathcal{S} \subseteq L_i$ are commensurable, i.e. $\mathcal{S} \times \mathcal{S} \subseteq K$.

Conversely, it can be shown that these conditions are also sufficient in order to characterize K as an abstract relation. In fact, in a lattice L_{qi} the relation K is *uniquely* defined by the conditions K1–K4. The proof is completely analogous to the proof of the corresponding theorem (2.18) about the commensurability relation in a quasimodular lattice L_q and will therefore not be repeated here.

If all elements of a lattice L_{qi} are pairwise commensurable, i.e. if we can add the relation $A \leq B \to A$ to the axioms, it follows by means of Lemma (5.24) that the lattice in question is implicative. This statement is, however, not sufficient in order to show that the lattice L_{qi} is only a weakening of the lattice L_i. In addition, one has to demonstrate that all axioms of L_{qi} are satisfied in an implicative lattice L_i. This can easily be done.

(5.26) THEOREM: The axioms $L_{qi}(1.1–5.3)$ of the lattice L_{qi} are satisfied in an implicative lattice L_i.

Proof: (a) The proofs of $L_{qi}(1.1–4.1)$ and of $L_{qi}(5.0–5.1)$ are obvious.

(b) $L_{qi}(4.2)$: From $A \wedge C \leq B$ it follows that $A \wedge (A \rightarrow C) \leq B$ and by means of $L_i(4.2)$ that $A \rightarrow C \leq A \rightarrow B$.

(c) $L_{qi}(5.2)$: From $A \wedge C \leq \Lambda$ it follows that $A \wedge (A \rightarrow C) \leq \Lambda$ and by means of $L_i(5.2)$ that $A \rightarrow C \leq \neg A$.

(d) $L_{qi}(4.3, 4.4, 5.3)$: If we put $C = B$, the premise of $L_i(4.2)$ is always fulfilled and thus for any two elements A, B the relation $B \leq A \rightarrow B$ holds.]

This theorem shows that the quasi-implicative lattice L_{qi} is, in fact, a weakening of the implicative L_i; this circumstance also justifies the name 'quasi-implicative' used here.

5.4 THE RELATION BETWEEN L_{qi} AND THE LATTICE L_q

In this section, we investigate the relation between the lattice L_{qi} of effective quantum logic and the orthocomplemented quasi-modular lattice L_q, which has as a model the lattice of subspaces of Hilbert space. This lattice L_q has already been discussed in detail in Chapter 2. Similarly, as in the preceding section, we shall show here that L_{qi} is also a weakening of the lattice L_q.

The lattice L_q can be defined by the following axioms: In the lattice L_q there exists a zero element Λ and a unit element V such that for any $A \in L_q$

$L_q(4.0)$ $\Lambda \leq A, A \leq \mathrm{V}$.

Furthermore, there exists an automorphism $\Theta_\neg : L_q \rightarrow L_q$, denoted by $\Theta_\neg(A) = \neg A$, which satisfies the conditions

$L_q(4.1)$ $A \wedge \neg A \leq \Lambda$,

$L_q(4.2)$ $\mathrm{V} \leq A \vee \neg A$,

$L_q(4.3)$ $A = \neg\neg A$,

$L_q(4.4)$ $A \leq B \curvearrowright \neg B \leq \neg A$.

Hence, L_q is *orthocomplemented* and the element $\neg A$ is the orthocomplement of A. Furthermore, the lattice L_q is quasimodular, i.e. the condition

$$L_q(5) \qquad B \leq A, C \leq \neg A \curvearrowright A \wedge (B \vee C) \leq B$$

is fulfilled for arbitrary elements $A, B, C \in L_q$.

A comparison of these axioms of L_q with the axioms of the lattice L_{qi} then leads to the following results: If one adds to the axioms $L_{qi}(1.1–5.3)$ of L_{qi} the additional postulates

$$\mathsf{V} \leq A \vee \neg A, \tag{5.27}$$

$$A = \neg\neg A, \tag{5.28}$$

$$A \leq B \curvearrowright \neg B \leq \neg A, \tag{5.29}$$

one obtains an orthocomplemented and quasimodular lattice. In fact, since the postulates (5.27)–(5.29) are formally equivalent to the axioms $L_q(4.2)$–(4.4), the quasi-pseudocomplement $\neg A$ defined within the framework of L_{qi} then goes over into an orthocomplement. Moreover, according to Theorem (5.17) the 'weak quasi-modularity'

$$B \leq A, C \leq \neg A \curvearrowright A \wedge (B \vee C) = B$$

is already valid in the framework of the lattice L_{qi}, whereby $\neg A$ is the quasi-pseudocomplement. Consequently, since after the addition of (5.27)–(5.29) to the axioms of L_{qi} the quasi-pseudocomplement goes over into an orthocomplement, the quasimodular law $L_q(5)$ is also fulfilled in this lattice. Therefore, if the three postulates (5.27)–(5.29) are added to the axioms of L_{qi}, the resulting lattice is orthocomplemented and quasi-modular, i.e. all axioms of L_q are fulfilled.

If the lattice L_{qi} is extended in this way by the addition of (5.27)–(5.29), the *material quasi-implication* $A \rightarrow B$ defined within the lattice L_{qi} goes over into the element $\neg A \vee (A \wedge B)$, which – in the framework of L_q – has likewise been called *material quasi-implication*. This can be seen in the following way: According to $L_{qi}(4.1)$, $L_q(4.2)$ and the relation $\neg A \vee (A \wedge C) \leq A \rightarrow C$, which follows from Theorem (5.11a, 5.11b) within the lattice L_{qi}, the material quasi-implication $A \rightarrow B$ fulfills the conditions

$$A \wedge (A \rightarrow B) \leq B, \tag{5.30}$$

$$A \wedge C \leq B \curvearrowright \neg A \vee (A \wedge C) \leq A \rightarrow B. \tag{5.31}$$

Furthermore, if the lattice L_{qi} is extended by (5.27)–(5.29) all axioms of L_q are fulfilled in the resulting lattice. On the other hand, according to Theorems (2.28) and (2.33), within the lattice L_q, due to the conditions (5.30) and (5.31), the element $A \to B$ is uniquely defined and given by $A \to B = \neg A \vee (A \wedge B)$. Hence, in the extension of the lattice L_{qi} just mentioned, the *material quasi-implication* $A \to B$ equals the L_q*-material quasi-implication* $\neg A \vee (A \wedge B)$.

If one wants to extend the lattice L_{qi} by additional axioms such that the resulting lattice is *orthocomplemented* and *quasimodular* it is not necessary to require the three conditions (5.27)–(5.29) since (5.27) is already sufficient. Moreover, there are some other postulates which are interesting in a certain respect and which are equivalent to (5.27).

(5.32) THEOREM: In the quasi-implicative lattice L_{qi}, the following four additional postulates are equivalent.[9]

(α) $\quad \mathsf{V} \leq A \vee \neg A$,

(β) $\quad \mathsf{V} \leq A \vee (A \to B)$,

(γ) $\quad \neg\neg A \leq A$,

(δ) $\quad (A \to B) \to A \leq A$.

Remark: In the logical interpretation (α) is called 'tertium non datur', whereas (β) is a generalization of this law. Condition (γ) is called 'stability' and the condition (δ) is known as 'Peirce's law'. These four conditions are fulfilled in the lattice L_q (cf. Theorem (2.37)).

Proof: (1) $\alpha \curvearrowright \gamma$: From ($\alpha$) it follows that $\neg\neg A = \neg\neg A \wedge (A \vee \neg A)$. From Theorem (5.11d), i.e. $A \leq \neg\neg A$ and Theorem (5.16a) one obtains $\neg\neg A \leq A \to \neg\neg A$, and using L_{qi}(5.3) it follows $\neg\neg A \leq \neg A \to \neg\neg A$. Hence, we can apply Theorem (5.18) and obtain

$$\neg\neg A = (\neg\neg A \wedge A) \vee (\neg\neg A \wedge \neg A) \leq A.]\!]$$

(2) $\gamma \curvearrowright \delta$: From Theorem (5.10) and L_{qi}(4.3, 5.3) we obtain $\neg A \leq A \to \neg A$, $\neg A \leq (A \to B) \to \neg A$ and by means of L_{qi}(4.4) it follows that $\neg A \leq ((A \to B) \to A) \to \neg A$, and thus

$$(A \to B) \to A \leq \neg A \to ((A \to B) \to A). \tag{5.33}$$

From L_{qi}(4.2) we obtain $A \to B \leq A \to (A \wedge B)$ and $A \to B \leq A \to C$ if $B \leq C$. Applying these formulae to the right-hand side of (5.33), we obtain finally

$$(A\to B)\to A \leqslant \neg A\to(\neg A\wedge((A\to B)\to A))$$
$$\leqslant \neg A\to((A\to B)\wedge((A\to B)\to A))$$
$$\leqslant \neg A\to A\leqslant \neg A\to \Lambda \leqslant \neg\neg A.$$

Hence, $\neg\neg A\leqslant A$ implies $(A\to B)\to A\leqslant A$.]]

(3) $\delta \curvearrowright \beta$: If we replace, in the formula (δ), A by $A\vee(A\to B)$ and B by A we obtain

$$((A\vee(A\to B))\to A)\to(A\vee(A\to B))\leqslant A\vee(A\to B). \qquad (5.34)$$

From Theorem (5.10) and $L_{qi}(4.4)$ we obtain $A\leqslant(A\vee(A\to B))\to A$. To prove the inverse relation we first show that $(A\vee(A\to B))\to A\leqslant (A\to B)\to A$. From Theorem (5.10) and $L_{qi}(4.3)$ we obtain

$$(A\vee(A\to B))\to A\leqslant(A\to B)\to((A\vee(A\to B))\to A),$$

and using $L_{qi}(4.2)$ twice it follows that

$$(A\to B)\to((A\vee(A\to B))\to A)\leqslant(A\to B)\to A.$$

Applying again formula (δ), we thus obtain $(A\vee(A\to B))\to A=A$. If we insert this result into (5.34) we obtain finally

$$\mathsf{V}\leqslant A\to A\leqslant A\to(A\vee(A\to B))\leqslant A\vee(A\to B).]]$$

(4) $\beta \curvearrowright \alpha$: If we put $B=\Lambda$ in formula (β), (α) follows.]]
It is a consequence of Theorem (5.32) that (5.27) implies (5.28). Furthermore, on account of Theorem (5.16) the condition (5.28) implies (5.29). Hence, if one adds the 'tertium non datur' law $\mathsf{V}\leqslant A\vee\neg A$ to the axioms of L_{qi}, one obtains an orthocomplemented and quasimodular lattice.

Consequently, the axioms $L_q(1.1–4.4)$ of the lattice L_q are fulfilled in the extension of the quasi-implicative lattice L_{qi} by the additional axiom $\mathsf{V}\leqslant A\vee\neg A$. Conversely, it can also be shown that in an orthocomplemented, quasimodular lattice L_q all axioms of L_{qi} are satisfied.

(5.35) THEOREM: In the lattice L_q the axioms $L_{qi}(1.1–5.3)$ are fulfilled.

Proof: $L_{qi}(1.1–3.3)$ are equivalent to $L_q(1.1–3.3)$. $L_{qi}(4.1)$ and $L_{qi}(5.1)$ are equivalent to $L_q(4.1)$ and $L_q(5.1)$, respectively. $L_{qi}(5.0)$ follows from $L_q(4.0)$.

$L_{qi}(4.2)$: From $A \to C = \neg A \vee (A \to C)$ and (5.31) it follows that $A \wedge C \leqslant B \curvearrowright A \to C \leqslant A \to B$.

$L_q(4.3)$: $A \leqslant B \to A$ implies $A \leqslant \neg B \vee (A \wedge B)$, and by means of de Morgan's formulae it follows that $B \wedge \neg(A \wedge B) \leqslant \neg A$. Applying (2.8) to $(A \wedge B) \vee (B \wedge \neg(A \wedge B))$, we obtain finally $B \leqslant \neg A \vee (A \wedge B)$.

$L_{qi}(4.4)$: From the premises we obtain $B \leqslant \neg A \vee (A \wedge B)$ and $C \leqslant \neg A \vee (A \wedge C)$ and hence $B \wedge C \leqslant \neg A \vee (A \wedge (B \wedge C))$ and $B \vee C \leqslant \neg A \vee (A \wedge (B \vee C))$. $B \to C$ can be eliminated by $\neg B \vee (B \wedge C)$.

$L_{qi}(5.2)$: From $A \to C = \neg A \vee (A \wedge C)$ and (5.31) it follows that $A \wedge C \leqslant \Lambda \curvearrowright A \to C \leqslant A \to \Lambda = \neg A$.

$L_{qi}(5.3)$: $A \leqslant \neg B \vee (A \wedge B)$ implies $B \wedge \neg(A \wedge B) \leqslant \neg A \wedge B$. Applying (2.8) to $\neg B \vee (B \wedge \neg(A \wedge B))$ we obtain $\neg A \leqslant \neg B \vee (\neg A \wedge B)$.]

Using Theorems (5.32) and (5.35), we thus arrive at the result that the lattice L_{qi} is in fact a weakening of the lattice L_q from which it differs only by the *tertium non datur* law $V \leqslant A \vee \neg A$.

As a conclusion of Sections (5.3) and (5.4), we have the following

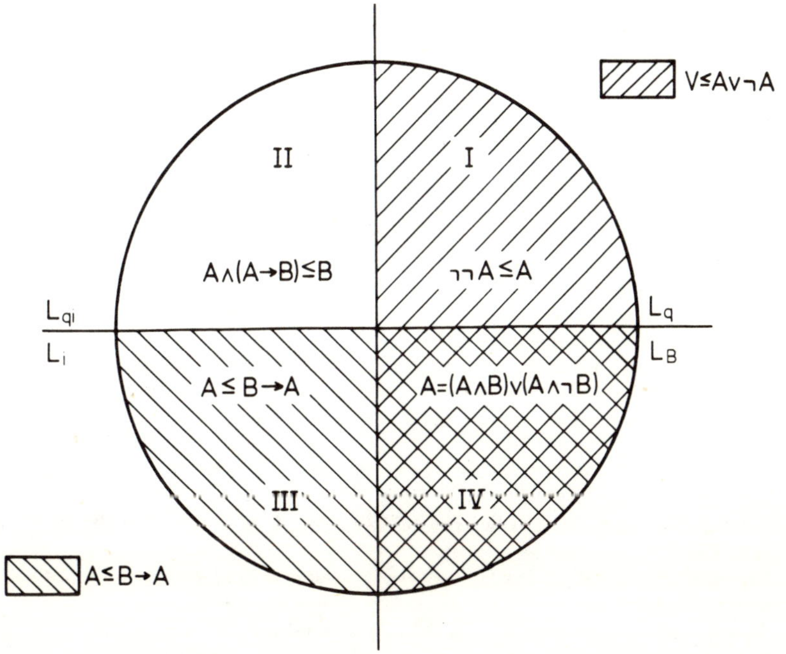

Fig. 5.1. The relations among the lattices L_{qi}, L_i, L_q and L_B.

relations among the lattices L_{qi}, L_q, L_i and L_B:

L_{qi} extended by $A \leqslant B \rightarrow A$ results in L_i
L_{qi} extended by $V \leqslant A \vee \neg A$ results in L_q
L_q extended by $A \leqslant B \rightarrow A$ results in L_B
L_i extended by $V \leqslant A \vee \neg A$ results in L_B

These relations can easily be illustrated by considering the set Σ_B of all propositions $\alpha \leqslant \beta$ which can be derived in L_B. Clearly the sets Σ_{qi}, Σ_q, Σ_i of propositions derivable in L_{qi}, L_q, L_i respectively are subsets of Σ_B. Furthermore, Σ_{qi} is a subset of Σ_q as well as of Σ_i and $\Sigma_{qi} = \Sigma_q \cap \Sigma_i$.[10] In Fig. 5.1, the area of the circle corresponds to Σ_B. The quadrants I and II correspond to Σ_q, the quadrants II and III correspond to Σ_i. Hence, the quadrant II represents Σ_{qi}. The propositions in quadrants I and IV are derived by means of the law $V \leqslant A \vee \neg A$, whereas the propositions in quadrants III and IV are derived by means of the law $A \leqslant B \rightarrow A$. Quadrant IV contains those propositions which can only be derived by means of $V \leqslant A \vee \neg A$ *and* $A \leqslant B \rightarrow A$. To each quadrant, we have attached an example which is an element of the quadrant considered.

NOTES AND REFERENCES

[1] A comprehensive discussion of the notions of deducible and admissible rules in a calculus can be found in P. Lorenzen, *Formal Logic*, D. Reidel Publishing Co., Dordrecht, Holland, (1965), p. 40ff.

[2] Another example of a logical calculus which is syntactically complete is the Brower-calculus L_{eff} of effective logic. This calculus can be found in ref. 1, p. 56. The syntactical completeness of L_{eff} is an immediate consequence of the more general results obtained in ref. 1.

[3] E.W. Stachow, *J. Philos. Logic.* **7**, (1978).

[4] G. Birkhoff, *Lattice Theory*, 3rd edn., Am. Math. Soc. Publ. Vol. XXV, Providence, Rhode Island (1973).

[5] H.B. Curry, *Foundations of Mathematical Logic*, Chapter 5, McGraw-Hill Book Co., New York (1963).

[6] P. Mittelstaedt and E.W. Stachow, *Found. of Physics* **4** (1974) 355.

[7] E.W. Stachow, Diplomarbeit, Köln (1973).

[8] Details of the implicative lattice can be found in ref. 5, Chapter 5, "The theory of implication".

[9] The equivalence of these four postulates is known to hold in the implicative lattice L_i (cf. refs. 5 and 7).

[10] $\Sigma_{qi} \subseteq \Sigma_q$ and $\Sigma_{qi} \subseteq \Sigma_i$ imply $\Sigma_{qi} \subseteq \Sigma_q \cap \Sigma_i$. The inverse relation can be obtained by the following argument: For L_B one can use the axioms of L_{qi}, $A \leqslant B \rightarrow A$ and $V \leqslant A \vee \neg A$. Hence, a relation $\alpha \in \Sigma_i \cap \Sigma_q$ is deducible from L_{qi}(1.1–5.3). Therefore, we have $\Sigma_q \cap \Sigma_i \subseteq \Sigma_{qi}$.

CHAPTER 6

THE CALCULUS OF FULL QUANTUM LOGIC

In this chapter, we establish the calculus Q of full quantum logic which is obtained from Q_{eff} by addition of the principle of excluded middle $\mathsf{V} \leqslant A \vee \neg A$ for all propositions A. In Section 6.1, we investigate the value-definiteness of material propositions. Starting from the value-definiteness of elementary propositions we show that the availability propositions k and $\bar{k}$ are, in general, not value-definite. In Section 6.2, we demonstrate that in contrast to ordinary effective logic the value-definiteness of the elementary propositions in Q_{eff} is not inherited by finitely compound propositions. It turns out that the reason for the missing principle of excluded middle for compound propositions is the lack of value-definiteness of the commensurability propositions k and $\bar{k}$ (Section 6.3). If, however, by a rather weak assumption concerning the measurability of k and $\bar{k}$, the value-definiteness of the availability propositions is re-established, it follows that the principle of excluded middle which is valid for elementary propositions, is, in fact, inherited by all finitely compound propositions. In this way, the calculus Q of full quantum logic can be justified (Section 6.4).

6.1 VALUE-DEFINITE MATERIAL PROPOSITIONS

6.1.1 *Elementary Propositions*

An elementary proposition $a = a(S)$ which asserts that a system S has a certain measurable property has to be proved or disproved by an experiment. According to the discussion in Section (3.1), we assume here that an elementary proposition a is proof-definite as well as disproof-definite. Hence, a *counter-proposition* $\bar{a}$ can be defined such that

$$\vdash\!\!\!\top\, a \leftrightarrow\, \vdash \bar{a}, \qquad (6.1)$$
$$\vdash a \leftrightarrow\, \vdash\!\!\!\top\, \bar{a}.$$

Obviously $\bar{a}$ is also proof-definite and disproof-definite (cf. 3.1 and 3.2). It follows from this definition of $\bar{a}$ that a and $\bar{\bar{a}}$ are equivalent in this way.

Elementary propositions a and $\bar{a}$ can be proved or disproved within the material quantum dialog-game D_m according to the argument-rules for a and $\bar{a}$ and by dialog schemes of the kind (3.5). Hence, we obtain, in accordance with the definition (3.6), that

$$\begin{aligned} &a \text{ is } m\text{-true} \leftrightharpoons \vdash_{D_m} a \leftrightharpoons \vdash a \\ &\bar{a} \text{ is } m\text{-true} \leftrightharpoons \vdash_{D_m} \bar{a} \leftrightharpoons \vdash \bar{a} \end{aligned} \tag{6.2}$$

Furthermore, if we recall the dialog (3.13) about the negation $\neg a$ and the definition (3.7), we find

$$\begin{aligned} &a \text{ is } m\text{-false} \leftrightharpoons \vdash_{D_m} \neg a \leftrightharpoons \Vdash a \\ &\bar{a} \text{ is } m\text{-false} \leftrightharpoons \vdash_{D_m} \neg\bar{a} \leftrightharpoons \Vdash \bar{a}, \end{aligned} \tag{6.3}$$

and hence

$$\vdash_{D_m} \neg a \leftrightharpoons \vdash_{D_m} \bar{a}, \tag{6.4}$$

using Definition (6.1).

In addition to the assumptions just mentioned we shall assume here that for an elementary proposition a there exists either a proof or a disproof, that is, we have either $\vdash a$ or $\vdash \bar{a}$. Propositions which have this additional property are called *value-definite*. The assumption of the value-definiteness of the elementary propositions means that for every proposition of this kind there exists an experimental testing procedure which decides between truth and falsity of the respective proposition.[1]

From the value-definiteness of elementary propositions, it follows that within the framework of the material dialog-game D_m there is a strategy of success for the compound proposition $a \vee \bar{a}$. For the proof, we consider the dialog

	O	P
(0)	[ ]	$a \vee \bar{a}$
(1)	? (0)	$a\ \langle 0\rangle$
(2)	a? (1)	[ ]
(3)	[ ]	$\bar{a}\ \langle 0\rangle$
(4)	$\bar{a}$? (3)	$\bar{a}!\ \langle 3\rangle$

(6.5)

According to the assumed value-definiteness there exists either a proof for a or a disproof for a, which is equivalent to a proof for $\bar{a}$. Hence the proponent P has a strategy of success in Row $P(2)$ or in Row $P(4)$, respectively. Therefore P has a strategy of success for the compound proposition $a \vee \bar{a}$, i.e. we have

$$\vdash_{D_m} a \vee \bar{a}. \tag{6.6}$$

Conversely, if there is a strategy of success for $a \vee \bar{a}$ within the material dialog-game D_m the proposition a is value-definite. Therefore, the material property of value-definiteness of a is equivalent to the material truth of the proposition $a \vee \bar{a}$.

In the next step, we incorporate the value-definiteness of the elementary propositions into the formal quantum dialog-game. In the formal dialog-game D_f only the opponent can guarantee the material truth of an elementary proposition. Therefore, if the value-definiteness of elementary propositions is taken into account in the framework of D_f, the opponent must guarantee also this material property of elementary propositions. Since the value-definiteness of a proposition a is equivalent to the material truth of $a \vee \bar{a}$, P should always be allowed to refer to $a \vee \bar{a}$ as a proposition of O. This can easily be achieved if all propositions $a_i \vee \bar{a}_i$ $(i = 1, \ldots N)$ are presupposed as hypotheses in Rows $O(-1)$, $O(-2) \ldots O(-N)$ of the dialog.

Furthermore, there is no system S for which the counter-proposition a and $\bar{a}$ can be proved at the same time (cf. Section 3.1). It is obvious that this material property of elementary propositions must also be guaranteed by the opponent. Therefore we postulate that O is not allowed to state $\bar{a}$ if he has asserted a previously and a is still available and he is not allowed to state a if he has asserted $\bar{a}$ previously and $\bar{a}$ is still available.

If these additional rules are introduced into the formal quantum dialog-game D_f, it can be shown that $\bar{a}$ and $\neg a$ are dialog-equivalent. Two propositions A and B are said to be *dialog-equivalent* – we write $A =_D B$ – if and only if they can be interchanged in any dialog without changing the strategic possibilities O and P. For the proof of the dialog-equivalence of $\bar{a}$ and $\neg a$ we consider in (6.7) a position in which P is allowed to state $\bar{a}$. This is the case only if he can take over $\bar{a}$ from O. In this situation, $\bar{a}$ can obviously be replaced by $\neg a$. In fact, since O has asserted $\bar{a}$ in Row $O(m)$ he is not allowed to attack $\neg a$ by a.

	O	P
$\vdots$	$\vdots$	
(m)	$\bar{a}$	
$\vdots$	$\vdots$	
(n)	$\vdots$	$\bar{a} \mid \neg a$
$(n+1)$	$\vdots$	

(6.7)

Conversely $\neg a$ can always be eliminated in the dialog if – as in (6.8) – O presupposes the hypothesis $a \vee \bar{a}$.

	O	P
(-1)	$a \vee \bar{a}$	
$\vdots$		
(n)		$\neg a \mid ?(-1)$
$(n+1)$	$a(n) \mid a\langle -1\rangle \mid \bar{a}\langle -1\rangle$	$[\ ]$

(6.8)

Instead of asserting $\neg a$ in $P(n)$, P attacks $a \vee \bar{a}$ in Row $O(-1)$. O can then defend in $O(n+1)$ either by asserting a or by asserting $\bar{a}$. In either case, there is no change in the further course of the dialog. Hence, we obtain the desired result

$$\bar{a} =_D \neg a, \tag{6.9}$$

which is, of course, a strengthening of (6.4). Using this dialog-equivalence (6.9) the counter-proposition $\bar{a}$ of an elementary proposition a can now be completely replaced by the negation $\neg a$. Consequently, we find, on account of (6.6), that the value-definiteness of an elementary proposition a is equivalent to the material truth of $a \vee \neg a$, i.e.

$$\vdash_{D_m} a \vee \neg a. \tag{6.10}$$

The proposition $a \vee \neg a$ is called the *tertium non datur* law or the *principle of excluded middle*, the validity of which has been shown here for elementary propositions.

6.1.2 *Commensurability Propositions*

The availability propositions $k(A, B)$ and $\bar{k}(A, B)$ have been introduced by the definitions (3.14) and (3.15). According to these definitions, the commensurability $k(A, B)$ and the incommensurability

$\bar{k}(A, B)$ are counter-propositions, i.e. (cf. (3.16))

$$\vdash k(A, B) \curvearrowright \Vdash \bar{k}(A, B) \qquad \vdash \bar{k}(A, B) \curvearrowright \Vdash k(A, B).$$

It follows from the definitions (3.14) and (3.15) that, for the demonstration of $\vdash k(A, B)$ and $\Vdash \bar{k}(A, B)$, infinite sequences of alternating dialogs about A and B are required. (If A and B are elementary propositions, infinite series of experimental tests are needed.) If one denotes the outcome of the nth pair of dialogs about A and B by $d_n(A, B)$ then one can express $\vdash k(A, B)$ by

$$\vdash k(A, B) \curvearrowright \underset{i=1}{\forall} \{d_i(A, B) = d_1(A, B)\}, \tag{6.11}$$

where $\forall_i$ is the infinite quantifier 'for all i'. The material proposition $k(A, B)$ must be '*verified*' like an experimental proposition, and this obviously requires an infinite number of steps.[4]

On the other hand, it also follows from the definitions (3.14) and (3.15) that the demonstration of $\Vdash k(A, B)$ and $\vdash \bar{k}(A, B)$ can be performed by a finite number of steps. In fact, for the disproof of $k(A, B)$ one has only to show that there is at least one dialogic outcome $d_n(A, B)$ which differs from $d_1(A, B)$, i.e.

$$\Vdash k(A, B) \curvearrowright \underset{i}{\exists} \{d_i(A, B) \neq d_1(A, B)\} \tag{6.12}$$

where $\exists_i$ is the existence quantifier. Obviously the disproof of $k(A, B)$ corresponds to the '*falsification*' of an experimental proposition, which can always be performed by finite means.

The infinite number of steps which is required in the proof of $k(A, B)$ and in the disproof of $\bar{k}(A, B)$ has the consequence that the value-definiteness of $k(A, B)$ and $\bar{k}(A, B)$ cannot be legitimized. In fact, there is no reason to assume that for any system and for arbitrary propositions A and B there exists either a proof of $k(A, B)$ (i.e. $\vdash k(A, B)$) or a disproof (i.e. $\vdash \bar{k}(A, B)$). Instead, it is obvious that there is no *finite* experimental and (or) dialogical testing procedure which decides between the truth and the falsity of the commensurability propositions $k(A, B)$ and $\bar{k}(A, B)$.

It is a consequence of this lack of value-definiteness that in contrast to the situation for elementary propositions (cf. (6.6)) the proposition $k(A, B) \vee \bar{k}(A, B)$ cannot be proved within the framework of the material dialog-game D_m. Furthermore, and for the same reason, we cannot prove the dialog-equivalence of $\bar{k}(A, B)$ and $\neg k(A, B)$ (i.e. $\bar{k}(A, B) =_D \neg k(A, B)$). However, there are two weakenings of this relation which can be proved, also without assuming the value-definiteness of $k(A, B)$. Firstly, it follows from the definitions of $k(A, B)$ and $\bar{k}(A, B)$ in the same way as (6.4) that

$$\vdash_{D_m} \neg k(A, B) \leftrightarrow \vdash_{D_m} \bar{k}(A, B). \tag{6.13}$$

This relation means that P has a strategy against k if and only if he has a strategy of success for $\bar{k}(A, B)$. Furthermore, it can be shown that $\neg\bar{k}(A, B)$ and $\neg\neg k(A, B)$ are dialog-equivalent, which is, of course, a weakening of the relation $\bar{k}(A, B) =_D \neg k(A, B)$ mentioned above.

For the proof of $\neg\bar{k}(A, B) =_D \neg\neg k(A, B)$, we consider the following dialog:

	O		P	
(n)			$\neg\bar{k}$	$\neg\neg k$
$(n+1)$	$\bar{k}(n)$	$\neg k(n)$	[ ]	[ ]
$(n+2)$		[ ]	$\bar{k}?(n+1)$	$k(n)$
$(n+3)$		$k?(n+2)$		

(6.14)

It follows from the definition of k and $\bar{k}$, that P has a strategy against $\bar{k}$ (in the left dialog) if and only if he has a strategy for k (in the right dialog). In both cases P puts the empty argument in Row $(n+1)$. Hence, if P has no strategy against $\bar{k}$ and no strategy for k and if there are still obligations of defence previous to Row n, he has (in both dialogs) to perform his last obligation first. Thus, in both situations, the strategic possibilities are equivalent, i.e. we have

$$\neg\bar{k}(A, B) =_D \neg\neg k(A, B). \tag{6.15}$$

Finally, it should be mentioned that in case the value-definiteness of k and $\bar{k}$ could be legitimized for some reason, the results (6.6), (6.9) and (6.10) could be derived in an analogous way for k and $\bar{k}$, i.e. in this case we would get $\vdash_{D_m} k \vee \bar{k}$, $\bar{k} =_D \neg k$ and $\vdash_{D_m} k \vee \neg k$, respectively.

6.2 THE VALUE-DEFINITENESS OF COMPOUND PROPOSITIONS

Compound propositions $A \in S$ or $A \in S^*$ are, in general, not proof-definite but dialog-definite. According to the definition (3.20) a proposition A is said to be *materially true* if and only if P has a strategy of success for A in D_m^* – and a proposition A is said to be *materially false* if and only if P has a strategy of success for $\neg A$ in D_m^*. Keeping these definitions in mind, we can now extend the concept of value-definiteness to compound propositions.

(6.16) DEFINITION: A proposition $A \in S$ or $A \in S^*$ is said to be *value-definite* if and only if the material dialog-game D_m or D_m^*, respectively, decides between the material truth and the material falsity of A.

It is obvious that, for the special case of proof-definite elementary propositions, Definition (6.16) is reduced to the definition of value-definiteness mentioned in Section 6.1. Furthermore, by means of a dialog which is analogous to (6.5) one can easily show that the value-definiteness of a compound proposition A is equivalent to the *material truth* of $A \vee \neg A$. Hence the value-definiteness of a certain proposition A could be again incorporated into the formal dialog-game by presupposing $A \vee \neg A$ as an hypothesis.

Elementary propositions $a \in S_e$ have been assumed to be value-definite. In the framework of the formal dialog-game, this means that for all propositions $a, b, c \in S_e$ the propositions $a \vee \neg a$, $b \vee \neg b$, $c \vee \neg c, \ldots$, respectively, have to be presupposed by the opponent as hypotheses in the dialog. However, it can easily be seen that starting from the hypotheses $a \vee \neg a$ and $b \vee \neg b$ there is no strategy of success within D_f for the principle of excluded middle for the connectives $a \wedge b$, $a \vee b$, $a \to b$, i.e. the propositions $(a * b) \vee \neg(a * b)$, $* \in \{\wedge, \vee, \to\}$ cannot be defended successfully by P. In the framework of the calculus Q_{eff} this means that the rule

$$\mathsf{V} \leq A \vee \neg A,, \mathsf{V} \leq B \vee \neg B \Rightarrow \mathsf{V} \leq (A * B) \vee \neg(A * B) \qquad (6.17)$$

cannot be deduced from $Q_{\text{eff}}(1.1\text{–}5.3)$ – or equivalently that in the quasi-implicative lattice L_{qi} the proposition

$$\mathsf{V} \leq A \vee \neg A, \mathsf{V} \leq B \vee \neg B \curvearrowright \mathsf{V} \leq (A * B) \vee \neg(A * B) \qquad (6.17a)$$

cannot be derived. Thus, within the framework of Q_{eff} the value-definiteness of the elementary proposition is not transferred to finitely compound propositions.

Before we discuss the meaning of this important statement, we shall briefly recall the situation in the calculus of ordinary effective logic L_{eff}, which can be represented by an implicative lattice L_i with zero element. The axioms of the lattice L_i have been formulated in Section 5.3. It is well-known that in the framework of the calculus L_{eff} the value-definiteness of elementary propositions is inherited by all finitely compound propositions.[5] In fact, we have

(6.18) THEOREM: In an implicative lattice L_i with zero element the following relations hold

$$\mathsf{V} \leq A \vee \neg A, \mathsf{V} \leq B \vee \neg B \curvearrowright \mathsf{V} \leq (A * B) \vee \neg(A * B) \quad \text{with } * \in \{\wedge, \vee, \rightarrow\},$$

$$\mathsf{V} \leq A \vee \neg A \curvearrowright \mathsf{V} \leq (\neg A) \vee \neg(\neg A).$$

Proof: The lattice L_i is distributive. Therefore, we obtain from the premises

$$\mathsf{V} \leq (A \wedge B) \vee (A \wedge \neg B) \vee (\neg A \wedge B) \vee (\neg A \wedge \neg B). \tag{6.19}$$

Starting from this relation, it follows that:

(a) From $A \wedge B \leq A \wedge B$, $A \wedge \neg B \leq \neg(A \wedge B)$, $\neg A \wedge B \leq \neg(A \wedge B)$, and $\neg A \wedge \neg B \leq \neg(A \wedge B)$ we obtain $\mathsf{V} \leq (A \wedge B) \vee \neg(A \wedge B)$.

(b) From $A \wedge B \leq A \rightarrow B$, $A \wedge \neg B \leq \neg(A \rightarrow B)$, $\neg A \wedge B \leq A \rightarrow B$, and $\neg A \wedge \neg B \leq A \rightarrow B$ we obtain $\mathsf{V} \leq (A \rightarrow B) \vee \neg(A \rightarrow B)$.

(c) From $A \wedge B \leq A \vee B$, $A \wedge \neg B \leq A \vee B$, $\neg A \wedge B \leq A \vee B$, and $\neg A \wedge \neg B \leq \neg(A \vee B)$ we obtain $\mathsf{V} \leq (A \vee B) \vee \neg(A \vee B)$.

(d) From $\mathsf{V} \leq A \vee \neg A$ we obtain $A = \neg\neg A$ and hence $\mathsf{V} \leq (\neg A) \vee \neg(\neg A)$.⟧

It is obvious that this proof cannot be transferred to the lattice L_{qi}. Since L_{qi} is not distributive, (6.19) cannot be derived. Moreover, for value-definite propositions A and B, (6.19) can be shown to be equivalent to $\mathsf{V} \leq k(A, B)$, i.e. to the commensurability of A and B (cf. (6.27)).

We will now come back to the above-mentioned statement that the proposition (6.17a) cannot be derived in L_{qi}. In order to illustrate this underivability in the most simple way, we consider the following situation: Let a and b be value-definite elementary propositions with possible truth-values 1 (true) and 0 (false). The truth-value of a compound proposition $a * b$ will then be defined to be 1 or 0 if P wins or loses the material dialog about $a * b$, respectively. For example, from the material dialog about $a \wedge b$ (cf. (3.17)) and the argument rule A_m(4a) one thus derives the following 'truth-table':

	O	P
(0)	[ ]	$a \wedge b$
(1)	1?	a
(2)	a?	a!
(3)	2?	b
(4)	b?	b!
(5)	$k(a, b)$?	

a	b	$a \wedge b$
1	1	(?)
1	0	0
0	1	0
0	0	0

(6.20)

If a and b have both the truth values 1, the truth-value of $a \wedge b$ is not determined since it still depends on the truth or falsity of the material proposition $k(a, b)$. It would be determined if in addition to the elementary propositions the commensurability proposition $k(a, b)$ were value-definite as well. This is, however, not the case. Since the material proposition $k(a, b)$ is defined by an infinite proof procedure, it will not have definite truth-values even if the elementary propositions a and b are value-definite. It should, however, be noticed that in the case that proposition $k(a, b)$ is value-definite, the truth value of $a \wedge b$ would not be determined by merely the truth values of a and b, but only by the truth values of a, b *and* $k(a, b)$. This is an immediate consequence of the argument rules $A_m(4a)$.

For the connectives $a \vee b$ and $a \rightarrow b$ the situation is quite similar. Since $k(a, b)$ and $\bar{k}(a, b)$ are not value-definite even for value-definite propositions a and b, the value definiteness of a and b is not inherited by the connectives $a \vee b$ and $a \rightarrow b$. Only the negation $\neg a$ is value-definite if a itself is value-definite. Thus, we arrive at the following result which is most remarkable: *If the elementary propositions are value-definite, this value-definiteness is not even inherited by finitely compound propositions.* From ordinary effective (intuitionistic) logic it is well-known that starting from value-definite elementary propositions the infinitely compound propositions (i.e. the quantifiers) can no longer be shown to be value-definite. However – as we have seen in Theorem (6.18) – all finitely compound propositions of L_i are, in fact, value-definite.

Thus, we have shown in this section that the value-definiteness of elementary propositions a and b is, in general, not inherited by the connectives $a \wedge b$, $a \vee b$ and $a \rightarrow b$, and this is so since the truth-value of $k(a, b)$ and $\bar{k}(a, b)$ is in general undetermined. However, the argument that the lack of value-definiteness of $k(a, b)$ is the entire reason for the missing principle of excluded middle for finitely compound propositions is for the present merely heuristic. For a conclusive justification of this argument one has to show that under the assumption that $k(a, b)$ and $\bar{k}(a, b)$ are value-definite, all finitely compound propositions are value-definite as well. This will, in fact, be demonstrated in the next section.

6.3 THE EXTENSION OF THE CALCULUS Q_{eff}

In order to justify in a formal manner the conjecture that the value-definiteness of the commensurabilities $k(A, B)$ and $\bar{k}(A, B)$ implies the value-definiteness of all finitely compound propositions, we have first to incorporate the material propositions $k(A, B)$ and $\bar{k}(A, B)$ into the calculus Q_{eff}. In order to do this we first extend the material dialog-game D_m and the formal dialog-game D_f to the dialog-games D_m^* and D_f^*, respectively, by incorporating the material propositions $k(A, B)$ and $\bar{k}(A, B)$. These extensions, which were already briefly mentioned in Sections 3.4 and 4.1, will be formulated in detail. In the next step, we establish the extended calculus Q_{eff}^* of effective quantum logic, which is again consistent and complete with respect to the extended formal quantum dialog-game D_f^*.[6]

We start from the extended set S^* of propositions which, apart from the logical connectives, also contains the commensurabilities $k(A, B)$ and $\bar{k}(A, B)$ as 2-place operations. Propositions $A \in S^*$ have to be proved within the framework of the material quantum dialog-game D_m^*. Those propositions $A \in S^*$, which can be shown to be true irrespective of the material propositions contained in them, are said to be formally true. An example of this kind is the proposition $A \equiv (a \wedge k(a, b)) \rightarrow (b \rightarrow a)$ which has been discussed in the dialog (4.2). The proof-procedure for these formally true propositions is then given by the extended formal quantum dialog-game D_f^*. The difference between D_f and D_f^* consists in the fact that commensurability propositions $k(A, B)$ and $\bar{k}(A, B)$ are still contained in D_f^* and that they are treated like elementary propositions. In particular, propositions $k(A, B)$ and $\bar{k}(A, B)$ are not attackable. On the other hand O is allowed to state $k(A, B)$ and $\bar{k}(A, B)$ in every position of the dialog, whereas P is allowed to state $k(A, B)$ and $\bar{k}(A, B)$ only if he can take over these propositions from O.

Furthermore, we take into account that P may take over the commensurability proposition $k(A, B)$ in every position of a dialog in which the conjunction $A \wedge B$ of O is available. This possibility is justified by the argument rule $A_m(4a)$. In addition, it follows from $A_m(4b)$ that P has a strategy of success for $A \vee B$ if the incommensurability proposition $\bar{k}(A, B)$ is available as a previous proposition of O. This possibility of defence will also be taken account of in the extended dialog-game D_f^*. In the formal dialog-game D_f this possibility was not used since in this game P has no strategy of success for $\bar{k}(A, B)$.

These additional possibilities of defence and attack in the dialog-game D_f^* can be summarized in the following reformulations of the argument-rules $A_f(1)$ and $A_f(4^{(n)}a)$:

$A_f^*(1)$(a): Material propositions are not attackable.

(b): O is allowed to state material propositions in every position of the dialog.

P is allowed to state material propositions only if they have been asserted by O previously and if they are still available – except $k(A, B)$ which he may also assert in case the conjunction $A \wedge B$ is available as a proposition of O.

(c): On an attack against the disjunction $A \vee B$, P is allowed to defend by asserting $\bar{k}(A, B)$ – provided that the conditions of $A_f^*(1b)$ are fulfilled.

$A_f^*(4^{(n)})$(a): P is allowed to assert the same material proposition in the case of $A_f^*(1b)$ at most n times. P is allowed to attack the same proposition of O at most n times.

By means of the argument-rules $A_f(1)$–$A_f(7)$, with the reformulations $A_f^*(1)$ and $A_f^*(4^{(n)})$, the formal dialog-game D_f^* is established. Similarly, as in Section 4.4, we can now use this dialog-game D_f^* in order to formulate the *extended calculus* Q_{eff}^* *of effective quantum logic*. It is obvious that all the beginnings and rules of Q_{eff} are also beginnings and rules of Q_{eff}^*. However, since the commensurability propositions $k(A, B)$ and $\bar{k}(A, B)$ are now incorporated into Q_{eff}^* we must add to Q_{eff} some beginnings and rules which govern the use of these propositions $k(A, B)$ and $\bar{k}(A, B)$ within the calculus Q_{eff}^*.

For the formulation of the additional beginnings and rules of Q_{eff}^* we use all the formal tools which were introduced in Section 4.4 and prove the new beginnings and rules by means of D_f^*-dialogs.

(6.21) THEOREM: $A \wedge B \leq k(A, B)$.

Proof: P can defend $k(A, B)$ in $P(2)$ by referring to $A \wedge B$ in $O(1)$

and using the rule A_f^*(1b).]]

	O	P
(0)	[ ]	$(A \wedge B) \rightarrow k(A, B)$
(1)	$A \wedge B$ (0)	$k(A, B)$ ⟨0⟩
(2)	$k(A; B)?$ (1)	$k(A, B)!$ ⟨1⟩

(6.22) THEOREM: $\bar{k}(A, B) \leq A \vee B$.

Proof: P can defend $A \vee B$ in $P(2)$ by taking over $\bar{k}(A, B)$ from $O(1)$ and using the rule A_f^*(1c).

	O	P
(0)	[ ]	$\bar{k}(A, B) \rightarrow (A \vee B)$
(1)	$\bar{k}(A, B)$ (0)	$A \vee B$ ⟨0⟩
(2)	? (1)	$\bar{k}(A, B)$ ⟨1⟩

In the dialog (6.14) we have shown that the propositions $\neg\bar{k}(A, B)$ and $\neg\neg k(A, B)$ are dialog-equivalent in the material dialog-game. Hence in the formal dialog-game D_f^* the material implications $\neg\bar{k}(A, B) \rightarrow \neg\neg k(A, B)$ and $\neg\neg k(A, B) \rightarrow \neg\bar{k}(A, B)$ can be successfully defended and thus we obtain the new beginning

$$\neg\bar{k}(A, B) = \neg\neg k(A, B) \tag{6.23}$$

of Q_{eff}^*.

If the propositions A and B are value-definite, the commensurability proposition $k(A, B)$ is dialog-equivalent to the compound proposition $k_0(A, B) \equiv (A \wedge B) \vee (A \wedge \neg B) \vee (\neg A \wedge B) \vee (\neg A \wedge \neg B)$ in the material dialog-game.[7] This can be demonstrated by a comparison of the dialogs for $k(A, B)$ and $k_0(A, B)$, respectively, using the definition (6.16) for the value-definiteness.[5] The series of dialogs about A and B, which establishes the proof of $k(A, B)$, has to be performed outside of the dialog-game which is governed by the frame-rules. Nevertheless, we may write the proof procedure for $k(A, B)$ in a dialog-like scheme. This is done in the schemes (6.24) and (6.25). In the first case, (6.24), it is assumed that P has a strategy of success for A and B (denoted by $\vdash_D A$ and $\vdash_D B$).

The comparison with the respective dialog (6.24)* for $k_0(A, B)$ shows that the possibilities for O and P in the two dialogs are, in fact, equivalent.

O	P
[ ]	$k(A, B)$
$k(A, B)$?	
- - - - - - -	- - - - - - -
[ ]	A
A?	$\vdash_D A$
- - - - - - -	- - - - - - -
[ ]	B
B?	$\vdash_D B$
- - - - - - -	- - - - - - -
[ ]	A
⋮	⋮

(6.24)

O	P
(0) [ ]	$k_0(A, B)$
(1) ? (0)	$A \wedge B\ \langle 0 \rangle$
- - - - - - -	- - - - - - -
(2) 1? (1)	$A\ \langle 1 \rangle$
(3) A? (2)	$\vdash_D A\ \langle 2 \rangle$
- - - - - - -	- - - - - - -
(4) 2? (1)	$B\ \langle 1 \rangle$
(5) B? (4)	$\vdash_D B\ \langle 4 \rangle$
- - - - - - -	- - - - - - -
(6) 1? (1)	$A\ \langle 1 \rangle$

(6.24*)

In the second case (6.25) it is assumed that P has a strategy of success for A but not for B. Since B is value-definite, this means, according to (6.9), that B is false and P has a strategy of success for $\neg B$ (denoted by $\vdash_D \neg B$). It is obvious that this case corresponds to the situation in the dialog (6.25*) where P has a strategy of success for $\neg B$. Analogously, it can be seen that the two remaining cases – where P has a strategy of success for B but not for A and where P has a strategy of success neither for A nor for B – correspond to dialogs about the other two conjunctions, $\neg A \wedge B$ and $\neg A \wedge \neg B$, of the proposition $k_0(A, B)$. Since the value-definiteness of a proposition A is equivalent to the material truth of $A \vee \neg A$, we have thus obtained the following result:

O	P
[ ]	$k(A, B)$
$k(A, B)?$	
---	---
[ ]	A
$A?$	$\vdash_{\overline{D}} A$
---	---
[ ]	B
$B?$	$\vdash_{\overline{D}} \neg B$

(6.25)

	O	P
(0)	[ ]	$k_0(A, B)$
(1)	? (0)	$A \wedge \neg B \langle 0 \rangle$
	---	---
(2)	1? (1)	$A \langle 1 \rangle$
(3)	A? (2)	$\vdash_{\overline{D}} A \langle 2 \rangle$
	---	---
(4)	2? (1)	$\neg B \langle 1 \rangle$
(5)	B (4)	[ ]
(6)	$\vdash_{\overline{D}} \neg B$	B? (5)

(6.25*)

(6.26) THEOREM: $\vdash_{\overline{D_m}} A \vee \neg A, \vdash_{\overline{D_m}} B \vee \neg B \curvearrowright k(A, B) =_D k_0(A, B)$.

In the formal dialog-game D_f^* the dialog-equivalence of $k(A, B)$ and $k_0(A, B)$ implies that the material implications $k(A, B) \rightarrow k_0(A, B)$ and $k_0(A, B) \rightarrow k(A, B)$ can successfully be defended. Thus, we obtain the new rule

$$\mathsf{V} \leqslant A \vee \neg A,, \mathsf{V} \leqslant B \vee \neg B \Rightarrow k(A, B) = k_0(A, B) \qquad (6.27)$$

of the extended calculus Q^*_{eff}.

If we recall (6.21), (6.22), (6.23) and (6.27), we can now formulate the beginnings and rules, which in addition to Q_{eff}(1.1–5.3) constitute the calculus Q^*_{eff}:

Q^*_{eff}(6.1) $A \wedge B \leqslant k(A, B)$,

Q^*_{eff}(6.2) $\bar{k}(A, B) \leqslant A \vee B$,

$Q^*_{\text{eff}}(6.3)$ $\neg\bar{k}(A, B) = \neg\neg k(A, B)$,

$Q^*_{\text{eff}}(6.4)$ $\mathsf{V} \leqslant A \vee \neg A,, \mathsf{V} \leqslant B \vee \neg B \Rightarrow k(A, B) = k_0(A, B)$.

This extended calculus Q^*_{eff} can again be shown to be consistent and complete with respect to the extended formal dialog-game D^*_f.

If one assumes that the propositions $k(A, B)$ and $\bar{k}(A, B)$ are value-definite, it follows by means of Q^*_{eff} that the proposition $\bar{k}(A, B)$ can be eliminated by $\neg k(A, B)$. In fact, we have

(6.28) THEOREM: In the calculus Q^*_{eff}, the rule

$$\mathsf{V} \leqslant k(A, B) \vee \bar{k}(A, B) \Rightarrow \bar{k}(A, B) = \neg k(A, B)$$

is deducible.

Remark: This theorem corresponds to the fact that for value-definite k and $\bar{k}$ it follows that $\bar{k}(A, B) =_D \neg k(A, B)$ (cf. (6.9)).

Proof: From the premise and $k(A, B) \leqslant \neg\neg k(A, B)$ and $\neg\bar{k}(A, B) = \neg\neg k(A, B)$ it follows that $\mathsf{V} \leqslant \neg\bar{k}(A, B) \vee \bar{k}(A, B)$. Thus, we get by means of Theorem (5.32) $\neg\neg\bar{k}(A, B) = \bar{k}(A, B)$ and again using (6.23), $\bar{k}(A, B) = \neg\neg\bar{k}(A, B) = \neg k(A, B)$.⟧

It is an immediate consequence of Theorem (6.28) that, under the assumption of the value-definiteness of $k(A, B)$, the *principle of excluded middle*

$$\mathsf{V} \leqslant k(A, B) \vee \neg k(A, B)$$

is valid for $k(A, B)$.

Finally, we mention a simple connection between formal commensurabilities and implications in Q^*_{eff}. If $k(A, B)$ is a formal commensurability which can be derived in the calculus K, we write $\vdash_K k(A, B)$. This is the case if and only if there is a strategy of success for $k(A, B)$ within the extended formal dialog-game D^*_f, i.e. if $\vdash_{D^*_f} k(A, B)$. Using the generalisation of the relation (4.17) for D^*_f we thus obtain

$$\vdash_K k(A, B) \curvearrowright \mathsf{V} \leqslant k(A, B). \tag{6.29}$$

In Q_{eff} the formal commensurability $k(A, B)$ is expressed by $A \leqslant B \rightarrow A$. In the extended calculus Q^*_{eff} formal commensurabilities can also be expressed by means of (6.29) using the material proposition $k(A, B)$.

6.4 THE PRINCIPLE OF EXCLUDED MIDDLE

In the framework of the calculus Q^*_{eff} we can now verify the conjecture mentioned above that under the assumption that the commensurability proposition $k(A, B)$ is value-definite the value-definiteness of two propositions A and B is inherited by all finitely compound propositions. According to Theorem (6.28) in the calculus Q_{eff} the value-definiteness of the proposition $k(A, B)$ means that the principle of excluded middle, i.e. $\mathsf{V} \leqslant k(A, B) \vee \neg k(A, B)$, is valid for $k(A, B)$. Hence, we arrive at the following result:

(6.30) THEOREM: In the calculus Q^*_{eff} the rules $\mathsf{V} \leqslant A \vee \neg A$,, $\mathsf{V} \leqslant B \vee \neg B$,, $\mathsf{V} \leqslant k(A, B) \vee \neg k(A, B) \Rightarrow \mathsf{V} \leqslant (A * B) \vee \neg(A * B)$ with $* \in \{\wedge, \vee, \rightarrow\}$ and $\mathsf{V} \leqslant A \vee \neg A \Rightarrow \mathsf{V} \leqslant (\neg A) \vee \neg(\neg A)$ can be deduced.

Remark: Theorem (6.30) is an obvious weakening of Theorem (6.18), which has been shown to hold in the calculus of ordinary effective logic which corresponds to the implicative lattice L_i. The weakening consists in the fact that the conclusion is stated here under the additional premise that for value-definite propositions A and B the relation $\mathsf{V} \leqslant k(A, B) \vee \neg k(A, B)$ holds.

Proof: (a) $\mathsf{V} \leqslant (A \wedge B) \vee \neg(A \wedge B)$.
From the premises we obtain, by means of Theorem (5.18),

$$\begin{aligned} \mathsf{V} &\leqslant (A \vee \neg A) \wedge (k(A, B) \vee \neg k(A, B)) \\ &= (A \wedge k(A, B)) \vee (A \wedge \neg k(A, B)) \vee (\neg A \wedge k(A, B)) \\ &\quad \vee (\neg A \wedge \neg k(A, B)) \end{aligned} \tag{6.31}$$

using the commensurability of A and $k(A, B)$. From the premises and $Q^*_{\text{eff}}(6.4)$ it follows that $k(A, B) = k_0(A, B)$ and hence (again using (5.18))

$$\begin{aligned} A \wedge k(A, B) &\leqslant (A \wedge B) \vee (A \wedge \neg B) \\ &\leqslant (A \wedge B) \vee \neg(A \wedge B), \end{aligned} \tag{α}$$

$$\begin{aligned} \neg A \wedge k(A, B) &\leqslant (\neg A \wedge B) \vee (\neg A \wedge \neg B) \\ &\leqslant (A \wedge B) \vee \neg(A \wedge B). \end{aligned} \tag{β}$$

From $Q^*_{\text{eff}}(6.1)$ we get $\neg k(A, B) \leqslant \neg(A \wedge B)$, and thus

$$A \wedge \neg k(A, B) \leqslant (A \wedge B) \vee \neg(A \wedge B), \tag{γ}$$

$$\neg A \wedge \neg k(A, B) \leqslant (A \wedge B) \vee \neg(A \wedge B). \tag{δ}$$

Inserting (α), (β), (γ), (δ) into (6.31) we finally obtain $\mathsf{V} \leq (A \wedge B) \vee \neg(A \wedge B)$.]]

(b) $\mathsf{V} \leq (A \vee B) \vee \neg(A \vee B)$.

From the premises we obtain (6.31). The four terms in the disjunction can then be shown to be below $(A \vee B) \vee \neg(A \vee B)$. In fact, it is

$$A \wedge k(A, B) \leq A \vee B \leq (A \vee B) \vee \neg(A \vee B) \qquad (\alpha^*)$$

$$\neg A \wedge k(A, B) \leq k(A, B) \leq (A \vee B) \vee \neg(A \vee B). \qquad (\beta^*)$$

Here, we have used Theorems (5.18) and (6.25). Furthermore, by means of $Q_{\text{eff}}(6.2)$ and Theorem (6.28) we obtain $\neg k(A, B) \leq A \vee B$ and hence

$$A \wedge \neg k(A, B) \leq (A \vee B) \vee \neg(A \vee B), \qquad (\gamma^*)$$

$$\neg A \wedge \neg k(A, B) \leq (A \vee B) \vee \neg(A \vee B). \qquad (\delta^*)$$

Inserting (α^*), (β^*), (γ^*) and (δ^*) into (6.31) we thus obtain $\mathsf{V} \leq (A \vee B) \vee \neg(A \vee B)$.]]

(c) $\mathsf{V} \leq (\neg A) \vee \neg(\neg A)$.

From the premise $\mathsf{V} \leq A \vee \neg A$ and Theorem (5.32) we obtain $A = \neg\neg A$ and hence $\mathsf{V} \leq (\neg A) \vee \neg(\neg A)$. This deduction is equivalent to the proof of Theorem (6.18d).

(d) $\mathsf{V} \leq (A \to B) \vee \neg(A \to B)$.[8]

For value-definite propositions A and B the material implication $A \to B$ can be expressed in the framework of Q^*_{eff} by the other connectives due to $A \to B = \neg A \vee (A \wedge B)$. In fact, from $\mathsf{V} \leq A \vee \neg A$, it follows that $A \to B = (A \vee \neg A) \wedge (A \to B)$. Using Theorem (5.18) and the formal commensurabilities $\mathsf{V} \leq k(A, A \to B)$ and $\mathsf{V} \leq k(\neg A, A \to B)$ we obtain

$$A \to B = (A \wedge (A \to B)) \vee (\neg A \wedge (A \to B)),$$

and on account of $A \wedge (A \to B) = A \wedge B$ and $\neg A \leq A \to B$ we obtain finally

$$A \to B = \neg A \vee (A \wedge B). \qquad (6.32)$$

From the proofs (a), (b), (c) already performed it follows that $A \wedge B$, $A \vee B$ and $\neg A$ are value-definite for value-definite propositions A and B. Applying these results to $\neg A \vee (A \wedge B)$ and using the equivalence (6.32) we obtain $\mathsf{V} \leq (A \to B) \vee \neg(A \to B)$.]]

It follows from Theorem (6.30) that starting from value-definite propositions $A, B, C, \ldots$ all finitely compound propositions are value-definite as well – provided that the commensurability proposition $k(A, B)$ of value-definite propositions A and B is itself value-definite. If this hypothesis concerning the commensurability proposition could be justified, the value-definiteness of the elementary propositions would imply that in Q^*_{eff} the principle of excluded middle $\mathsf{V} \leqslant A \vee \neg A$ is valid for all finitely compound propositions. In this case, the calculus Q^*_{eff} could be extended by the incorporation of the principle of excluded middle as an additional beginning.

However, it is still necessary to justify the hypothesis that for value-definite propositions A and B the commensurability propositions $k(A, B)$ and $\bar{k}(A, B)$ are value-definite also. This can be done by the following reasoning. The proof procedures of $k(A, B)$ and $\bar{k}(A, B)$, which are performed outside of the dialog, require an infinite number of steps. This has been expressed in (6.11) and (6.12) by means of quantifiers. Since, however, infinite proof procedures cannot be performed in practice, the participants P and O of the dialog may introduce the following strategy in order to establish the commensurability of two propositions: They postulate that $k(A, B)$ and $\bar{k}(A, B)$ – as material propositions – can be confirmed by a sufficiently large but finite number of dialogs. Formally this *strategy of confirmation* (C-strategy) means that the material quantifier proposition (6.11) is replaced by a conjunction of arbitrary but finite length.

If once this C-strategy is accepted one can argue in the following way: If two propositions A and B are value-definite, any finite series of proofs will itself be value-definite. Therefore, if A and B are value-definite, it follows that $k(A, B)$ is also value-definite and a proof procedure decides between $k(A, B)$ and $\bar{k}(A, B)$. Consequently, if the participants are using the C-strategy, the calculus Q^*_{eff} can be extended by the additional rule

$$Q^*_{\text{eff}}(C) \qquad \mathsf{V} \leqslant A \vee \neg A,, \mathsf{V} \leqslant B \vee \neg B \Rightarrow \mathsf{V} \leqslant k(A, B) \vee \bar{k}(A, B)$$

from which we immediately obtain

$$\mathsf{V} \leqslant A \vee \neg A,, \mathsf{V} \leqslant B \vee \neg B \Rightarrow \mathsf{V} \leqslant k(A, B) \vee \neg k(A, B) \qquad (6.33)$$

using Theorem (6.28).

The acceptance of the C-strategy corresponds to the assumption that a material proposition which contains quantifiers can be decided in a finite number of steps. This hypothesis is generally accepted in physics in as much as the fundamental laws of nature, which always contain quantifiers, are considered to be justifiable by a finite verification procedure – in spite of the well-known problem of induction.[9] Moreover, in the classical logic of quantifiers – which is mostly used in mathematics – the principle of excluded middle is even stated as a formal law, although the intuitionistic critique of this assumption is still unchallenged.[10] Here, however, we state only the weaker hypothesis, that the material quantifier propositions $k(A, B)$ and $\bar{k}(A, B)$ are value-definite. Hence, accepting the C-strategy, we make an assumption which is indeed presently unproved but which is questioned seriously neither in physics nor in mathematics.

If for these reasons the rule (6.33) is taken for granted we can extend the calculus Q^*_{eff} by the additional rule $Q^*_{\text{eff}}(C)$. In the framework of this extended calculus $(Q^*_{\text{eff}}, Q^*_{\text{eff}}(C))$ the value-definiteness of two propositions A and B is, in fact, inherited by all finitely compound propositions. This is an immediate consequence of Theorem (6.30), which can thus be reformulated in the following way:

(6.30*) THEOREM: In the calculus $(Q^*_{\text{eff}}, Q^*_{\text{eff}}(C))$ the rules

$$\mathsf{V} \leqslant A \vee \neg A, \mathsf{V} \leqslant B \vee \neg B \Rightarrow \mathsf{V} \leqslant (A * B) \vee \neg(A * B)$$
$$\mathsf{V} \leqslant A \vee \neg A \Rightarrow \mathsf{V} \leqslant (\neg A) \vee \neg(\neg A)$$

can be deduced.

Elementary propositions $a, b, c, \dots$ are value-definite. This value-definiteness can be expressed in Q^*_{eff} by the relations $\mathsf{V} \leqslant a \vee \neg a$, $\mathsf{V} \leqslant b \vee \neg b$, $\mathsf{V} \leqslant c \vee \neg c, \dots$, respectively. In order to incorporate the value-definiteness of the elementary propositions into the calculus $(Q^*_{\text{eff}}, Q^*_{\text{eff}}(C))$ we extend this calculus by the additional beginnings $\mathsf{V} \leqslant a \vee \neg a$ for all elementary propositions a. The resulting calculus will be designated by Q^{**}_{eff}. It then follows by means of Theorem (6.30*) that in the calculus Q^{**}_{eff} the principle of excluded middle, that is the relation $\mathsf{V} \leqslant A \vee \neg A$, is valid for all finitely compound propositions.

In the calculus Q^{**}_{eff} there are still contained the material propositions $k(A, B)$ and $\bar{k}(A, B)$. According to $Q^*_{\text{eff}}(C)$ these propositions are value-definite. Furthermore, from the value-definiteness of $k(A, B)$ it follows by means of Theorem (6.28) that $\bar{k}(A, B) = \neg k(A, B)$ and hence the incommensurability proposition $\bar{k}(A, B)$ can be eliminated in Q_{eff}^{**}. Moreover, for value-definite propositions A and B it follows from $Q_{\text{eff}}^{*}(6.4)$ that $k(A, B)$ can be expressed by the compound proposition $k_0(A, B)$ which is given by (6.25). Therefore, the commensurability proposition $k(A, B)$ can also be eliminated in Q_{eff}^{**}. The calculus which is obtained from Q_{eff}^{**} by elimination of $k(A, B)$ and $\bar{k}(A, B)$ agrees exactly with the calculus Q, which will be called the *calculus of full quantum logic* and which is defined in the following way:

(6.34) DEFINITION: The calculus Q of full quantum logic is given by the calculus Q_{eff} and the principle of excluded middle $\mathsf{V} \leqslant A \vee \neg A$ as an additional beginning.

Thus we find that the extension of the calculus of effective quantum logic Q_{eff} by the principle of excluded middle $\mathsf{V} \leqslant A \vee \neg A$ can, in fact, be justified if for the availability propositions $k(A, B)$ and $\bar{k}(A, B)$ the C-strategy is accepted. The resulting calculus Q of full quantum logic can – under this presupposition – be considered as the logical calculus of arbitrary quantum mechanical propositions. On the other hand, the calculus Q of full quantum logic is – according to the investigations in Section 5.4 – a model of the orthocomplemented quasimodular lattice L_q. Thus, we arrive at the important result that the lattice L_q can in fact be interpreted as the logic of quantum-physical propositions.

The lattice L_q was obtained in Chapter 2 from the lattice L_H of subspaces of a Hilbert space by leaving out the more special property of atomicity and the covering law. Since, on the other hand, the calculus Q of full quantum logic has been established here by a priori reasoning only, it follows that at least the algebraic structure of a Hilbert space, which is given by the lattice L_q, can be justified without any recourse to empirical knowledge.

NOTES AND REFERENCES

[1] Within the framework of ordinary logic the problem of the value-definiteness of elementary propositions is discussed, e.g. in ref. 2. In quantum logic, the more general case of elementary propositions, which are not necessarily value-definite, is treated in ref. 3.

[2] W. Kamlah and P. Lorenzen, *Logische Propädeutik*, Bibliographisches Institut, Mannheim (1967/73).

[3] E.W. Stachow, 'Quantum logical calculi', *J. Philos. Logic* **7** (1978).

[4] The problems of verification and falsification are discussed by K. Popper, 'The Logic of Scientific Discovery', London (1959). Cf. also Note 9.

[5] The fact that in the ordinary effective logic the value-definiteness of two propositions is inherited by the connectives $\wedge$, $\vee$, $\rightarrow$ and $\neg$ is discussed in refs. 2 and 6.

[6] P. Mittelstaedt and E.W. Stachow, 'The principle of excluded middle in quantum logic', *J. Philos. Logic* **7** (1978) 181.

[7] The proposition $k_0(A, B)$ has already been introduced within the framework of the lattice L_q as a commensurability proposition. In fact, in L_q we have $\mathsf{V} = k_0(A, B)$ iff $A \sim B$. However, in the calculus Q^*_{eff} one has clearly to distinguish between the commensurability proposition $k(A, B)$, which cannot be expressed by the other connectives, and the compound proposition $k_0(A, B)$.

[8] A more extensive proof of this relation, which does not make use of the possibility of replacing $A \rightarrow B$ by the other connectives, can be found in ref. 6.

[9] It has been emphasized in particular by K. Popper, *The Logic of Scientific Discovery*, London (1959), that the fundamental laws of physics can only be 'falsified' but not 'verified'. An attempt to justify the verification of physical laws – in spite of Popper's critique – was made by R. Carnap, *Induktive Logik und Wahrscheinlichkeit*, Wien (1959).

[10] The problem of the justification of the 'tertium non datur' in the logic of quantifiers is treated in detail in P. Lorenzen, *Formal Logic*, D. Reidel Publishing Co., Dordrecht, Holland, (1965), p. 86ff.

CONCLUDING REMARKS: CLASSICAL LOGIC AND QUANTUM LOGIC

The considerations of Chapters 3–5 have led to the result that the most general propositional logic which is equally applicable to propositions about classical and quantum mechanical systems is given by the calculus Q_{eff} of effective quantum logic. In addition, in Chapter 6 we have shown that, by a weak assumption concerning the confirmation of commensurability propositions, this calculus can be extended to the calculus Q of full quantum logic. Hence, one arrives at the conclusion that the 'true' logic is given by the calculus of full quantum logic, whereas ordinary effective logic and classical propositional logic are formal systems, the validity of which is restricted to the special case of unrestrictedly available (commensurable) propositions. In particular, propositions about physical systems have this property to the extent that the system in question can be considered as a classical one. Consequently, from this point of view classical propositional logic turns out to be an idealisation, which has only approximate validity and, thus, no fundamental meaning.

This argument is confirmed in some sense by the following observation: There is still another connection between quantum logic and classical logic which shows that the ordinary effective logic is indeed an indispensible but not an independent constituent of the scientific language of physics. This connection becomes obvious if, in addition to the *object-logic* of propositions about physical systems, the *meta-logic* of quantum logic is considered. By meta-logic, we mean the formal logic of those metapropositions which make assertions about the formal 'truth' or 'falsity' of an object-proposition within the framework of the effective (or full) quantum logic.

In Chapter 3, we developed the *object-language* of quantum physics which consists of elementary propositions, commensurabilities and compound propositions. These various kinds of propositions were defined by the possibilities of attacking and defending the respective proposition within the material and the formal dialog-game D_m and D_f, respectively. In particular, the logical connectives $\wedge$, $\vee$, $\rightarrow$ and $\neg$ were introduced by attack-and-defence rules which correspond to the dialogs (3.10–3.13). Starting from the formal dialog-game D_f, the concepts of formal 'truth' and 'falsity' of an arbitrarily compound proposition were then defined by $\vdash_{D_f} A$ and $\vdash_{D_f} \neg A$, respectively. The totality of all formally true propositions could eventually be comprehended by the calculus Q_{eff} of effective quantum logic and its extension Q, respectively. These calculi consist of statements $\alpha, \beta, \ldots$ about the formal truth of certain propositions (beginnings) and rules $\alpha \Rightarrow \beta$ which allow for the derivation of further statements of this kind.

On the basis of this quantum logic, one can now define *meta-propositions* 'A is formally true' by $\bar{A} \rightleftharpoons \vdash_{D_f} A$. These meta-propositons will be considered as elementary meta-propositions, the proof of which has to be performed by a D_f-dialog about A. In a succeeding step one can define meta-logical connectives by means of the possibilities of attack and defence in a *meta-dialog-game* $\bar{D}$. The dialogic definition of the meta-connectives, which will be denoted by $\bar{\wedge}$, $\underline{\vee}$, $\Rightarrow$ and $\dashv$, is in complete analogy to the definition of the connectives in the object-language, namely by the following attack-and-defence scheme:

Meta-connective	Attack	Defence
$\bar{A} \bar{\wedge} \bar{B}$	1?, 2?	$\bar{A}, \bar{B}$
$\bar{A} \underline{\vee} \bar{B}$	?	$\bar{A}, \bar{B}$
$A \Rightarrow B$	$\bar{A}$	$\bar{B}$
$\dashv A$	$\bar{A}$	

An elementary meta-proposition $\bar{A}$ can either be expressed in terms of dialogs by $\vdash_{D_f} A$ or using the calculus Q_{eff} by $\vdash_{Q_{\text{eff}}} \mathsf{V} \leq A$. In both formulations, it becomes evident that all elementary meta-propositions are mutually commensurable, i.e. $k(\bar{A}, \bar{B})$ is valid for all meta-propositions of this kind. In fact, in the first formulation, $\bar{A}$ asserts the existence of a strategy of success for A in D_f whereas, in the second formulation, $\bar{A}$ states that $\mathsf{V} \leq A$ can be deduced within the calculus Q_{eff}. It is a consequence of this commensurability property that in contrast to the dialogs (3.10–3.13) the respective meta-dialogs for the connectives of elementary meta-propositions can be restricted to a finite number of steps.

The meta-logical material implication has been denoted here by '$\Rightarrow$' which is the same sign as that for rules of the calculus Q_{eff}. For the present, one has clearly to distinguish between the meta-logical material implication $\vdash_{\bar{D}_f} A \Rightarrow \vdash_{\bar{D}_f} B$ and the respective rule $\mathsf{V} \leqslant A \Rightarrow \mathsf{V} \leqslant B$ in Q_{eff}. The precise connection between a material implication in $\bar{D}$ and a rule in Q_{eff} is as follows: The rule $\alpha \Rightarrow \beta$ is admissible in Q_{eff} if and only if there is a strategy of success for the material implication $\alpha \Rightarrow \beta$ in the meta-dialog-game $\bar{D}$. This relationship between the calculus Q_{eff} of the object-logic and strategies in the meta-dialog-game $\bar{D}$ can easily be illustrated by means of meta-dialogs. It is well-known that in the ordinary effective logic an analogous situation exists.

The formation of compound meta-propositions by means of the meta-connectives can easily be iterated by again using the meta-dialogic definition of $\bar{\wedge}$, $\underline{\vee}$, $\Rightarrow$ and $\dashv$. Since elementary meta-propositions are mutually commensurable and since the meta-connectives are defined by finite meta-dialogs, all compound meta-propositions containing one connective are commensurable as well. Hence, compound meta-propositions which contain more than one meta-connective can again be defined by meta-dialogs with a finite number of steps. In this way, one arrives at the important result that all finitely compound meta-propositions are mutually commensurable and thus unrestrictedly available in a meta-dialog.

Analogous to the situation described in Chapter 4 one can now define a formal meta-dialog-game $\bar{D}_f$ and formally true meta-propositions. The totality of all formally true meta-propositions will then constitute the formal *meta-logic of quantum logic*. Since meta-propositions are mutually commensurable one thus obtains the logic of unrestrictedly available propositions, i.e. the ordinary effective logic. Hence, one finds that the formal meta-logic of quantum logic is the effective logic. The principle of excluded middle can generally be proved here no more than in the well-known dialogic justification of ordinary logic.[1,2] The question whether the special type of elementary meta-propositions discussed here is value-definite and, thus, the principle of excluded middle can be justified for all finitely compound meta-propositions, is still open.

The relationship between quantum logic and classical logic which is obtained in this way can be summarized as follows: On the one hand the 'true' object-logic of propositions about physical systems is given by the calculus of full quantum logic. Classical logic appears as an idealisation which possesses only approximate validity. On the other hand, the meta-logic of quantum logic which is generated by the formalism of quantum logic itself agrees exactly with the ordinary effective logic. Quantum logic is not the meta-logic of quantum logic.[3,4,5] Both approaches demonstrate that quantum logic is in fact prior to classical logic which turns out to be either an approximation or a derived structure which is valid for the special case of meta-propositions.

NOTES AND REFERENCES

[1] K. Lorenz, 'Die dialogische Rechtfertigung der effektiven Logik' in: F. Kambartel und J. Mittelstraß (Eds), *Zum normativen Fundament der Wissenschaft*, Athenäum, Frankfurt (1973).

[2] P. Lorenzen, *Formal Logic*, D. Reidel Publishing Co., Dordrecht, Holland, (1965), p. 66ff.

[3] It is a much debated question whether quantum logic is also the meta-logic of quantum logic. Here, we mention in particular the contributions of H. Putnam[4] and P. Heelan.[5]

[4] H. Putnam, 'Is Logic Empirical?', *Boston Studies in the Philosophy of Science* **5**, D. Reidel Publishing Co., Dordrecht, Holland (1969), p. 199.

[5] P. Heelan, 'Quantum and Classical Logic: Their respective roles', *Synthese* **21** (1970) p. 1.

BIBLIOGRAPHY

This bibliography summarizes the literature quoted in the references and, in addition, contains books and articles which are of interest for the topic of this book. The bibliography does not claim to exhaust the field of quantum logic; it should, however, be sufficient for an adequate understanding of the problems treated in the present book.

Achieser, N.J. and Glasmann, J.M., *Theorie der linearen Operatoren im Hilbert-Raum*, Akademie Verlag, Berlin (1958).

Beltrametti, E.G. and Casinelli, G., 'On state transformations induced by yes-no experiments in the context of quantum logic', *J. Philos. Logic* **6** (1977) 369.

Birkhoff, G., *Lattice Theory*, third edn., American Mathematical Society, Providence, Rhode Island (1973).

Birkhoff, G. and v. Neumann, J., 'The logic of quantum mechanics', *Ann. of Math.* **37** (1936) 823.

Bub, J., *The Interpretation of Quantum Mechanics*, D. Reidel Publishing Co., Dordrecht, Holland (1974).

Bugajska, K. and Bugajski, S., The projection postulate in quantum logic, *Bull. Acad. Pol. Sci., Ser. Sci. Math, Astron. Phys.* **21** (1973a) 873–877.

Carnap, R., *Induktive Logik und Wahrscheinlichkeit*, Springer, Wien (1959).

Curry, H.B., *Foundations of Mathematical Logic*, McGraw-Hill, New York (1963), in particular, Chapter 5, 'The theory of implication'.

Denecke, H.M., 'Quantum logic of quantifiers', *J. Philos. Logic* **6** (1977) 405.

Drieschner, M., 'Is (quantum) logic empirical?', *J. Philos. Logic* **6** (1977) 415.

d'Espagnat, B., *Conceptions de la physique contemporaine*, Hermann, Paris (1965).

Finkelstein, D., *Quantum Logic*, Wiley-Interscience, New York (1976).

Finkelstein, D., Jauch, J.M., and Speiser, D., 'Foundations of quaternion mechanics', *J. Math. Phys.* **3** (1962) 207.

Finkelstein, D., Jauch, J.M., Schiminovich, S., and Speiser, D., 'Principles of general *Q* covariance', *J. Math. Phys.* **4** (1963) 788.

Foulis, D.J., A note on orthomodular lattices, *Fortugaliae Mathematica* **21** (1962) 65.

van Fraassen, B.C., 'Semantic analysis of quantum logic', in: C.A. Hooker (Ed.), *Contemporary Research in the Foundations and the Philosophy of Quantum Theory*, D. Reidel Publishing Co., Dordrecht, Holland (1973), p. 80.

Gleason, A.M., 'Measures on the closed sub-spaces of a Hilbert space', *J. Math. Mech.* **6** (1957) 885–893.

Greechie, R.J. and Gudder, S.P., 'Quantum Logics', in: C.A. Hooker (Ed.), *Contemporary Research in the Foundations and Philosophy of Quantum Mechanics*, D. Reidel Publishing Co., Dordrecht, Holland (1975) p. 143.

Halmos, P.R., *Introduction to Hilbert Space*, Chelsea Publishing Co., New York (1957).

Heelan, P., 'Quantum and classical logic: their respective roles', *Synthese* **21** (1970) 1.

Holland, S., 'The current interest in orthomodular lattices', in: S.C. Abbott (ed.), *Trends in Lattice Theory*, Van Nostrand Reinhold, New York (1970), p. 41ff.

Hooker, C.A. (Ed.), *The Logico-Algebraic Approach to Quantum Mechanics* I, D. Reidel Publishing Co., Dordrecht, Holland (1975).

Jammer, M., *The Philosophy of Quantum Mechanics*, Wiley, New York (1974).

Jauch, J.M., *Foundations of Quantum Mechanics*, Addison-Wesley, Reading, Mass. (1968).

Jauch, J.M. and Piron, C., 'Can hidden variables be excluded in quantum mechanics?', *Helv. Phys. Acta* **36** (1963) 827.

Jauch, J.M. and Piron, C., 'What is quantum logic?', in: P.G.O. Freund et al. (Eds.), *Quanta*, University of Chicago Press, Chicago (1970), p. 166.

Kägi-Romano, U., 'Quantum logic and generalized probability theory', *J. Philos. Logic* **6** (1977) 455.

Kamber, F., 'Die Struktur des Aussagenkalküls einer physikalischen Theorie', *Nachr. Akad. Wiss. Math. Phys. Kl.* **10**, Göttingen (1964) p. 103; English translation in: C.A. Hooker (Ed.), *The Logico-Algebraic Approach to Quantum Mechanics* I, D. Reidel Publishing Co., Dordrecht, Holland (1975), p. 221.

Kamber, F., Zweiwertige Wahrscheinlichkeits-funktionen auf orthokomplementären Verbänden, *Math. Ann.* **158** (1965) 158.

Kamlah, W. and P. Lorenzen, *Logische Propädeutik*, Bibliographisches Institut, Mannheim (1973).

Kröger, H., 'Zwerch-Assoziativität und verbandsähnliche Algebren', *Sitzungsber. Bayer. Akad. Wiss.*, München (1973).

Kunsemüller, H., 'Zur Axiomatik der Quantenlogik', *Philosophia naturalis* **8** (1964) 363.

Lorenz, K., 'Dialogspiele als semantische Grundlagen von Logigkalkülen' *Arch. Math. Logik Grundlagenforsch.* **11** (1968) 32, 73.

Lorenz, K., 'Die dialogische Rechtfertigung der effektiven Logik', in: F. Kambartel und J. Mittelstraß (Eds.), *Zum normativen Fundament der Wissenschaft*, Athenäum, Frankfurt (1973).

Lorenzen, P., *Metamathematik*, Bibliographisches Institut, Mannheim (1962).

Lorenzen, P., *Formal Logic*, D. Reidel Publishing Co., Dordrecht, Holland (1965).

Ludwig, G., *Grundlagen der Quantenmechanik*, Springer-Verlag, Berlin (1954).

Mackey, G.W., *Mathematical Foundations of Quantum Mechanics*, W.A. Benjamin, New York (1963).

Mittelstaedt, P., 'Untersuchungen zur Quantenlogik', *Sitzungsber. Bayer. Akad. Wiss.*, München (1959).

Mittelstaedt, P., 'Quantenlogische Interpretation orthokomplementären quasimodularer Verbände', *Z. naturforsch.* **25a** (1970) 1773.

Mittelstaedt, P., 'On the interpretation of the lattice of subspaces of the Hilbert space as a propositional calculus', *Z. Naturforsch.* **27a** (1972) 1358.

Mittelstaedt, P., 'Quantum Logic', in: R.S. Cohen et al. (Eds.), *PSA 1974*, D. Reidel Publishing Co., Dordrecht, Holland (1976) p. 501.

Mittelstaedt, P., 'On the applicability of the probability concept to quantum theory', in: Harper and Hooker (Eds.), *Foundations of Probability Theory, Statistical Inference and Statistical Theories of Science*, Vol. III, D. Reidel Publishing Co., Dordrecht, Holland (1976).

Mittelstaedt, P., *Philosophical Problems of Modern Physics*, D. Reidel Publishing Co., Dordrecht, Holland (1976).

Mittelstaedt, P., 'Time-dependent propositions and quantum logic', *J. Philos. Logic* **6** (1977) 463.

Mittelstaedt, P. and Stachow, E.W., 'Operational foundation of quantum logic', *Found. Phys.* **4** (1974) 355.

Mittelstaedt, P. and Stachow, E.W., 'The principle of excluded middle in quantum logic', *J. Philos. Logic* **7** (1978) 181.

Nakamura, M., 'The permutability in a certain orthocomplemented lattice', *Kodai Math. Ser.*, Rep. **9** (1957) 158.

v. Neumann, J., *Mathematical Foundations of Quantum Mechanics*, Princeton University Press, Princeton (1955).

Ochs, W., 'On the covering law in quantal proposition systems', *Commun. Math. Phys.* **25** (1972) 245–252.

Piron, C., 'Axiomatique quantique', *Helv. Phys. Acta* **37** (1964) 439.

Piron, C., *Foundations of Quantum Physics*, W.A. Benjamin, Reading, Mass (1976).

Piron, C., Varenna lectures (1977).

Piron, C., 'On the logic of quantum logic', *J. Philos. Logic* **6** (1977) 481.

Pool, J.C.T., 'Baer†-Semigroups and the logic of quantum mechanics', *Commun. Math. Phys.* **9** (1968) 118.

Pool, J.C.T., Semi-Modularity and the logic of quantum mechanics, *Commun. Math. Phys.* **9** (1968) 212.

Popper, K., *The Logic of Scientific Discovery*, Basic Books, London (1959).

Popper, K., 'Birkhoff and von Neumann's interpretation of quantum mechanics', *Nature* **219** (1969) 682.

Putnam, H., 'Is logic empirical?', *Boston Studies in the Philosophy of Science* **5**, D. Reidel Publishing Co., Dordrecht, Holland (1969) p. 199.

Reichenbach, H., Philosophical Foundations of Quantum Mechanics, University of California Press (1948).

Scheibe, E., *Die kontingenten Aussagen in der Physik*, Athenäum, Frankfurt (1964).

Scheibe, E., *The Logical Analysis of Quantum Mechanics*, Pergamon Press, Oxford (1973).

Scheibe, E., 'Popper and quantum logic' *Brit. J. Philos. Sci.* **25** (1974) 319.

Stachow, E.W., Diplomarbeit, Köln (1973).

Stachow, E.W., Dissertation, Köln (1975).

Stachow, E.W., 'Completeness of quantum logic', *J. Philos. Logic* **5** (1976) 237.

Stachow, E.W., 'How does quantum logic correspond to physical reality?', *J. Philos. Logic* **6** (1977) 485.

Stachow, E.W., 'Quantum logical calculi', *J. Philos. Logic* **7** (1978).

Varadarajan, V.S., *Geometry of Quantum Theory*; Vol. 1, Van Nostrand, Princeton, N.J. (1968); Vol. 2, Van Nostrand-Reinhold, New York, (1970); Now: Springer, Berlin/New York.

v. Weizsäcker, C.F., 'Komplementärität und Logik', *Naturwiss.* **42** (1955) 521, 545.

INDEX OF NAMES AND SUBJECTS

SYNTHESE LIBRARY

Studies in Epistemology, Logic, Methodology, and Philosophy of Science

Managing Editor:
JAAKKO HINTIKKA, (Academy of Finland, Stanford University and Florida State University)

Editors:
ROBERT S. COHEN (Boston University)
DONALD DAVIDSON (University of Chicago)
GABRIËL NUCHELMANS (University of Leyden)
WESLEY C. SALMON (University of Arizona)

1. J. M. Bocheński, *A Precis of Mathematical Logic.* 1959, X + 100 pp.
2. P. L. Guiraud, *Problèmes et méthodes de la statistique linguistique.* 1960, VI + 146 pp.
3. Hans Freudenthal (ed.), *The Concept and the Role of the Model in Mathematics and Natural and Social Sciences. Proceedings of a Colloquium held at Utrecht, The Netherlands, January 1960.* 1961, VI + 194 pp.
4. Evert W. Beth, *Formal Methods. An Introduction to Symbolic Logic and the Study of Effective Operations in Arithmetic and Logic.* 1962, XIV + 170 pp.
5. B. H. Kazemier and D. Vuysje (eds.), *Logic and Language. Studies Dedicated to Professor Rudolf Carnap on the Occasion of His Seventieth Birthday.* 1962, VI + 256 pp.
6. Marx W. Wartofsky (ed.), *Proceedings of the Boston Colloquium for the Philosophy of Science 1961-1962,* Boston Studies in the Philosophy of Science (ed. by Robert S. Cohen and Marx W. Wartofsky), Volume I. 1963, VIII + 212 pp.
7. A. A. Zinov'ev, *Philosophical Problems of Many-Valued Logic.* 1963, XIV + 155 pp.
8. Georges Gurvitch, *The Spectrum of Social Time.* 1964, XXVI + 152 pp.
9. Paul Lorenzen, *Formal Logic.* 1965, VIII + 123 pp.
10. Robert S. Cohen and Marx W. Wartofsky (eds.), *In Honor of Philipp Frank,* Boston Studies in the Philosophy of Science (ed. by Robert S. Cohen and Marx W. Wartofsky), Volume II. 1965, XXXIV + 475 pp.
11. Evert W. Beth, *Mathematical Thought. An Introduction to the Philosophy of Mathematics.* 1965, XII + 208 pp.
12. Evert W. Beth and Jean Piaget, *Mathematical Epistemology and Psychology.* 1966, XII + 326 pp.
13. Guido Küng, *Ontology and the Logistic Analysis of Language. An Enquiry into the Contemporary Views on Universals.* 1967, XI + 210 pp.
14. Robert S. Cohen and Marx W. Wartofsky (eds.), *Proceedings of the Boston Colloquium for the Philosophy of Science 1964-1966, in Memory of Norwood Russell Hanson,* Boston Studies in the Philosophy of Science (ed. by Robert S. Cohen and Marx W. Wartofsky), Volume III. 1967, XLIX + 489 pp.

15. C. D. Broad, *Induction, Probability, and Causation. Selected Papers.* 1968, XI + 296 pp.
16. Günther Patzig, *Aristotle's Theory of the Syllogism. A Logical-Philosophical Study of Book A of the Prior Analytics.* 1968, XVII + 215 pp.
17. Nicholas Rescher, *Topics in Philosophical Logic.* 1968, XIV + 347 pp.
18. Robert S. Cohen and Marx W. Wartofsky (eds.), *Proceedings of the Boston Colloquium for the Philosophy of Science 1966-1968,* Boston Studies in the Philosophy of Science (ed. by Robert S. Cohen and Marx W. Wartofsky), Volume IV. 1969, VIII + 537 pp.
19. Robert S. Cohen and Marx W. Wartofsky (eds.), *Proceedings of the Boston Colloquium for the Philosophy of Science 1966-1968,* Boston Studies in the Philosophy of Science (ed. by Robert S. Cohen and Marx W. Wartofsky), Volume V. 1969, VIII + 482 pp.
20. J.W. Davis, D. J. Hockney, and W. K. Wilson (eds.), *Philosophical Logic.* 1969, VIII + 277 pp.
21. D. Davidson and J. Hintikka (eds.), *Words and Objections: Essays on the Work of W. V. Quine.* 1969, VIII + 366 pp.
22. Patrick Suppes, *Studies in the Methodology and Foundations of Science. Selected Papers from 1911 to 1969.* 1969, XII + 473 pp.
23. Jaakko Hintikka, *Models for Modalities. Selected Essays.* 1969, IX + 220 pp.
24. Nicholas Rescher *et al.* (eds.), *Essays in Honor of Carl G. Hempel. A Tribute on the Occasion of His Sixty-Fifth Birthday.* 1969, VII + 272 pp.
25. P. V. Tavanec (ed.), *Problems of the Logic of Scientific Knowledge.* 1969, XII + 429 pp.
26. Marshall Swain (ed.), *Induction, Acceptance, and Rational Belief.* 1970, VII + 232 pp.
27. Robert S. Cohen and Raymond J. Seeger (eds.), *Ernst Mach: Physicist and Philosopher,* Boston Studies in the Philosophy of Science (ed. by Robert S. Cohen and Marx W. Wartofsky), Volume VI. 1970, VIII + 295 pp.
28. Jaakko Hintikka and Patrick Suppes, *Information and Inference.* 1970, X + 336 pp.
29. Karel Lambert, *Philosophical Problems in Logic. Some Recent Developments.* 1970, VII + 176 pp.
30. Rolf A. Eberle, *Nominalistic Systems.* 1970, IX + 217 pp.
31. Paul Weingartner and Gerhard Zecha (eds.), *Induction, Physics, and Ethics: Proceedings and Discussions of the 1968 Salzburg Colloquium in the Philosophy of Science.* 1970, X + 382 pp.
32. Evert W. Beth, *Aspects of Modern Logic.* 1970, XI + 176 pp.
33. Risto Hilpinen (ed.), *Deontic Logic: Introductory and Systematic Readings* 1971, VII + 182 pp.
34. Jean-Louis Krivine, *Introduction to Axiomatic Set Theory.* 1971, VII + 98 pp.
35. Joseph D. Sneed, *The Logical Structure of Mathematical Physics.* 1971, XV + 311 pp.
36. Carl R. Kordig, *The Justification of Scientific Change.* 1971, XIV + 119 pp.
37. Milič Čapek, *Bergson and Modern Physics,* Boston Studies in the Philosophy of Science (ed. by Robert S. Cohen and Marx W. Wartofsky), Volume VII. 1971, XV + 414 pp.

38. Norwood Russell Hanson, *What I Do Not Believe, and Other Essays* (ed. by Stephen Toulmin and Harry Woolf), 1971, XII + 390 pp.
39. Roger C. Buck and Robert S. Cohen (eds.), *PSA 1970. In Memory of Rudolf Carnap,* Boston Studies in the Philosophy of Science (ed. by Robert S. Cohen and Marx W. Wartofsky), Volume VIII. 1971, LXVI + 615 pp. Also available as paperback.
40. Donald Davidson and Gilbert Harman (eds.), *Semantics of Natural Language.* 1972, X + 769 pp. Also available as paperback.
41. Yehoshua Bar-Hillel (ed.), *Pragmatics of Natural Languages.* 1971, VII + 231 pp.
42. Sören Stenlund, *Combinators, λ-Terms and Proof Theory.* 1972, 184 pp.
43. Martin Strauss, *Modern Physics and Its Philosophy. Selected Papers in the Logic, History, and Philosophy of Science.* 1972, X + 297 pp.
44. Mario Bunge, *Method, Model and Matter.* 1973, VII + 196 pp.
45. Mario Bunge, *Philosophy of Physics.* 1973, IX + 248 pp.
46. A. A. Zinov'ev, *Foundations of the Logical Theory of Scientific Knowledge (Complex Logic),* Boston Studies in the Philosophy of Science (ed. by Robert S. Cohen and Marx W. Wartofsky), Volume IX. Revised and enlarged English edition with an appendix, by G. A. Smirnov, E. A. Sidorenka, A. M. Fedina, and L. A. Bobrova. 1973, XXII + 301 pp. Also available as paperback.
47. Ladislav Tondl, *Scientific Procedures,* Boston Studies in the Philosophy of Science (ed. by Robert S. Cohen and Marx W. Wartofsky), Volume X. 1973, XII + 268 pp. Also available as paperback.
48. Norwood Russell Hanson, *Constellations and Conjectures* (ed. by Willard C. Humphreys, Jr.). 1973, X + 282 pp.
49. K. J. J. Hintikka, J. M. E. Moravcsik, and P. Suppes (eds.), *Approaches to Natural Language. Proceedings of the 1970 Stanford Workshop on Grammar and Semantics.* 1973, VIII + 526 pp. Also available as paperback.
50. Mario Bunge (ed.), *Exact Philosophy – Problems, Tools, and Goals.* 1973, X + 214 pp.
51. Radu J. Bogdan and Ilkka Niiniluoto (eds.), *Logic, Language, and Probability. A Selection of Papers Contributed to Sections IV, VI, and XI of the Fourth International Congress for Logic, Methodology, and Philosophy of Science, Bucharest, September 1971.* 1973, X + 323 pp.
52. Glenn Pearce and Patrick Maynard (eds.), *Conceptual Change.* 1973, XII + 282 pp.
53. Ilkka Niiniluoto and Raimo Tuomela, *Theoretical Concepts and Hypothetico-Inductive Inference.* 1973, VII + 264 pp.
54. Roland Fraïssé, *Course of Mathematical Logic* – Volume 1: *Relation and Logical Formula.* 1973, XVI + 186 pp. Also available as paperback.
55. Adolf Grünbaum, *Philosophical Problems of Space and Time.* Second, enlarged edition, Boston Studies in the Philosophy of Science (ed. by Robert S. Cohen and Marx W. Wartofsky), Volume XII. 1973, XXIII + 884 pp. Also available as paperback.
56. Patrick Suppes (ed.), *Space, Time, and Geometry.* 1973, XI + 424 pp.
57. Hans Kelsen, *Essays in Legal and Moral Philosophy,* selected and introduced by Ota Weinberger. 1973, XXVIII + 300 pp.
58. R. J. Seeger and Robert S. Cohen (eds.), *Philosophical Foundations of Science. Proceedings of an AAAS Program, 1969,* Boston Studies in the Philosophy of

Science (ed. by Robert S. Cohen and Marx W. Wartofsky), Volume XI. 1974, X + 545 pp. Also available as paperback.

59. Robert S. Cohen and Marx W. Wartofsky (eds.), *Logical and Epistemological Studies in Contemporary Physics*, Boston Studies in the Philosophy of Science (ed. by Robert S. Cohen and Marx W. Wartofsky), Volume XIII. 1973, VIII + 462 pp. Also available as paperback.
60. Robert S. Cohen and Marx W. Wartofsky (eds.), *Methodological and Historical Essays in the Natural and Social Sciences. Proceedings of the Boston Colloquium for the Philosophy of Science 1969-1972*, Boston Studies in the Philosophy of Science (ed. by Robert S. Cohen and Marx W. Wartofsky), Volume XIV. 1974, VIII + 405 pp. Also available as paperback.
61. Robert S. Cohen, J. J. Stachel and Marx W. Wartofsky (eds.), *For Dirk Struik. Scientific, Historical and Political Essays in Honor of Dirk J. Struik*, Boston Studies in the Philosophy of Science (ed. by Robert S. Cohen and Marx W. Wartofsky), Volume XV. 1974, XXVII + 652 pp. Also available as paperback.
62. Kazimierz Ajdukiewicz, *Pragmatic Logic*, transl. from the Polish by Olgierd Wojtasiewicz. 1974, XV + 460 pp.
63. Sören Stenlund (ed.), *Logical Theory and Semantic Analysis. Essays Dedicated to Stig Kanger on His Fiftieth Birthday*. 1974, V + 217 pp.
64. Kenneth F. Schaffner and Robert S. Cohen (eds.), *Proceedings of the 1972 Biennial Meeting, Philosophy of Science Association*, Boston Studies in the Philosophy of Science (ed. by Robert S. Cohen and Marx W. Wartofsky), Volume XX. 1974, IX + 444 pp. Also available as paperback.
65. Henry E. Kyburg, Jr., *The Logical Foundations of Statistical Inference*. 1974, IX + 421 pp.
66. Marjorie Grene, *The Understanding of Nature: Essays in the Philosophy of Biology*, Boston Studies in the Philosophy of Science (ed. by Robert S. Cohen and Marx W. Wartofsky), Volume XXIII. 1974, XII + 360 pp. Also available as paperback.
67. Jan M. Broekman, *Structuralism: Moscow, Prague, Paris*. 1974, IX + 117 pp.
68. Norman Geschwind, *Selected Papers on Language and the Brain*, Boston Studies in the Philosophy of Science (ed. by Robert S. Cohen and Marx W. Wartofsky), Volume XVI. 1974, XII + 549 pp. Also available as paperback.
69. Roland Fraïssé, *Course of Mathematical Logic* – Volume 2: *Model Theory*. 1974, XIX + 192 pp.
70. Andrzej Grzegorczyk, *An Outline of Mathematical Logic. Fundamental Results and Notions Explained with All Details*. 1974, X + 596 pp.
71. Franz von Kutschera, *Philosophy of Language*. 1975, VII + 305 pp.
72. Juha Manninen and Raimo Tuomela (eds.), *Essays on Explanation and Understanding. Studies in the Foundations of Humanities and Social Sciences*. 1976, VII + 440 pp.
73. Jaakko Hintikka (ed.), *Rudolf Carnap, Logical Empiricist. Materials and Perspectives*. 1975, LXVIII + 400 pp.
74. Milič Čapek (ed.), *The Concepts of Space and Time. Their Structure and Their Development*, Boston Studies in the Philosophy of Science (ed. by Robert S. Cohen and Marx W. Wartofsky), Volume XXII. 1976, LVI + 570 pp. Also available as paperback.

75. Jaakko Hintikka and Unto Remes, *The Method of Analysis. Its Geometrical Origin and Its General Significance,* Boston Studies in the Philosophy of Science (ed. by Robert S. Cohen and Marx W. Wartofsky), Volume XXV. 1974, XVIII + 144 pp. Also available as paperback.
76. John Emery Murdoch and Edith Dudley Sylla, *The Cultural Context of Medieval Learning. Proceedings of the First International Colloquium on Philosophy, Science, and Theology in the Middle Ages – September 1973,* Boston Studies in the Philosophy of Science (ed. by Robert S. Cohen and Marx W. Wartofsky), Volume XXVI. 1975, X + 566 pp. Also available as paperback.
77. Stefan Amsterdamski, *Between Experience and Metaphysics. Philosophical Problems of the Evolution of Science,* Boston Studies in the Philosophy of Science (ed. by Robert S. Cohen and Marx W. Wartofsky), Volume XXXV. 1975, XVIII + 193 pp. Also available as paperback.
78. Patrick Suppes (ed.), *Logic and Probability in Quantum Mechanics.* 1976, XV + 541 pp.
79. Hermann von Helmholtz: *Epistemological Writings. The Paul Hertz/Moritz Schlick Centenary Edition of 1921 with Notes and Commentary by the Editors.* (Newly translated by Malcolm F. Lowe. Edited with an Introduction and Bibliography, by Robert S. Cohen and Yehuda Elkana), Boston Studies in the Philosophy of Science (ed. by Robert S. Cohen and Marx W. Wartofsky), Volume XXXVII. 1977, XXXVIII+204 pp. Also available as paperback.
80. Joseph Agassi, *Science in Flux,* Boston Studies in the Philosophy of Science (ed. by Robert S. Cohen and Marx W. Wartofsky), Volume XXVIII. 1975, XXVI + 553 pp. Also available as paperback.
81. Sandra G. Harding (ed.), *Can Theories Be Refuted? Essays on the Duhem-Quine Thesis.* 1976, XXI + 318 pp. Also available as paperback.
82. Stefan Nowak, *Methodology of Sociological Research: General Problems.* 1977, XVIII + 504 pp.
83. Jean Piaget, Jean-Blaise Grize, Alina Szeminska, and Vinh Bang, *Epistemology and Psychology of Functions,* Studies in Genetic Epistemology, Volume XXIII. 1977, XIV+205 pp.
84. Marjorie Grene and Everett Mendelsohn (eds.), *Topics in the Philosophy of Biology,* Boston Studies in the Philosophy of Science (ed. by Robert S. Cohen and Marx W. Wartofsky), Volume XXVII. 1976, XIII + 454 pp. Also available as paperback.
85. E. Fischbein, *The Intuitive Sources of Probabilistic Thinking in Children.* 1975, XIII + 204 pp.
86. Ernest W. Adams, *The Logic of Conditionals. An Application of Probability to Deductive Logic.* 1975, XIII + 156 pp.
87. Marian Przełęcki and Ryszard Wójcicki (eds.), *Twenty-Five Years of Logical Methodology in Poland.* 1977, VIII + 803 pp.
88. J. Topolski, *The Methodology of History.* 1976, X + 673 pp.
89. A. Kasher (ed.), *Language in Focus: Foundations, Methods and Systems. Essays Dedicated to Yehoshua Bar-Hillel,* Boston Studies in the Philosophy of Science (ed. by Robert S. Cohen and Marx W. Wartofsky), Volume XLIII. 1976, XXVIII + 679 pp. Also available as paperback.
90. Jaakko Hintikka, *The Intentions of Intentionality and Other New Models for Modalities.* 1975, XVIII + 262 pp. Also available as paperback.

91. Wolfgang Stegmüller, *Collected Papers on Epistemology, Philosophy of Science and History of Philosophy*, 2 Volumes, 1977, XXVII + 525 pp.
92. Dov M. Gabbay, *Investigations in Modal and Tense Logics with Applications to Problems in Philosophy and Linguistics.* 1976, XI + 306 pp.
93. Radu J. Bogdan, *Local Induction.* 1976, XIV + 340 pp.
94. Stefan Nowak, *Understanding and Prediction: Essays in the Methodology of Social and Behavioral Theories.* 1976, XIX + 482 pp.
95. Peter Mittelstaedt, *Philosophical Problems of Modern Physics*, Boston Studies in the Philosophy of Science (ed. by Robert S. Cohen and Marx W. Wartofsky), Volume XVIII. 1976, X + 211 pp. Also available as paperback.
96. Gerald Holton and William Blanpied (eds.), *Science and Its Public: The Changing Relationship*, Boston Studies in the Philosophy of Science (ed. by Robert S. Cohen and Marx W. Wartofsky), Volume XXXIII. 1976, XXV + 289 pp. Also available as paperback.
97. Myles Brand and Douglas Walton (eds.), *Action Theory. Proceedings of the Winnipeg Conference on Human Action, Held at Winnipeg, Manitoba, Canada, 9-11 May 1975.* 1976, VI + 345 pp.
98. Risto Hilpinen, *Knowledge and Rational Belief.* 1979 (forthcoming).
99. R. S. Cohen, P. K. Feyerabend, and M. W. Wartofsky (eds.), *Essays in Memory of Imre Lakatos*, Boston Studies in the Philosophy of Science (ed. by Robert S. Cohen and Marx W. Wartofsky), Volume XXXIX. 1976, XI + 762 pp. Also available as paperback.
100. R. S. Cohen and J. J. Stachel (eds.), *Selected Papers of Léon Rosenfeld*, Boston Studies in the Philosophy of Science (ed. by Robert S. Cohen and Marx W. Wartofsky), Volume XXI. 1978, XXX + 927 pp.
101. R. S. Cohen, C. A. Hooker, A. C. Michalos, and J. W. van Evra (eds.), *PSA 1974: Proceedings of the 1974 Biennial Meeting of the Philosophy of Science Association*, Boston Studies in the Philosophy of Science (ed. by Robert S. Cohen and Marx W. Wartofsky), Volume XXXII. 1976, XIII + 734 pp. Also available as paperback.
102. Yehuda Fried and Joseph Agassi, *Paranoia: A Study in Diagnosis*, Boston Studies in the Philosophy of Science (ed. by Robert S. Cohen and Marx W. Wartofsky), Volume L. 1976, XV + 212 pp. Also available as paperback.
103. Marian Przełęcki, Klemens Szaniawski, and Ryszard Wójcicki (eds.), *Formal Methods in the Methodology of Empirical Sciences.* 1976, 455 pp.
104. John M. Vickers, *Belief and Probability.* 1976, VIII + 202 pp.
105. Kurt H. Wolff, *Surrender and Catch: Experience and Inquiry Today*, Boston Studies in the Philosophy of Science (ed. by Robert S. Cohen and Marx W. Wartofsky), Volume LI. 1976, XII + 410 pp. Also available as paperback.
106. Karel Kosík, *Dialectics of the Concrete*, Boston Studies in the Philosophy of Science (ed. by Robert S. Cohen and Marx W. Wartofsky), Volume LII. 1976, VIII + 158 pp. Also available as paperback.
107. Nelson Goodman, *The Structure of Appearance*, Boston Studies in the Philosophy of Science (ed. by Robert S. Cohen and Marx W. Wartofsky), Volume LIII. 1977, L + 285 pp.
108. Jerzy Giedymin (ed.), *Kazimierz Ajdukiewicz: The Scientific World-Perspective and Other Essays, 1931 – 1963.* 1978, LIII + 378 pp.

109. Robert L. Causey, *Unity of Science.* 1977, VIII+185 pp.
110. Richard E. Grandy, *Advanced Logic for Applications.* 1977, XIV + 168 pp.
111. Robert P. McArthur, *Tense Logic.* 1976, VII + 84 pp.
112. Lars Lindahl, *Position and Change: A Study in Law and Logic.* 1977, IX + 299 pp.
113. Raimo Tuomela, *Dispositions.* 1978, X + 450 pp.
114. Herbert A. Simon, *Models of Discovery and Other Topics in the Methods of Science,* Boston Studies in the Philosophy of Science (ed. by Robert S. Cohen and Marx W. Wartofsky), Volume LIV. 1977, XX + 456 pp. Also available as paperback.
115. Roger D. Rosenkrantz, *Inference, Method and Decision.* 1977, XVI + 262 pp. Also available as paperback.
116. Raimo Tuomela, *Human Action and Its Explanation. A Study on the Philosophical Foundations of Psychology.* 1977, XII + 426 pp.
117. Morris Lazerowitz, *The Language of Philosophy. Freud and Wittgenstein,* Boston Studies in the Philosophy of Science (ed. by Robert S. Cohen and Marx W. Wartofsky), Volume LV. 1977, XVI + 209 pp.
118. Tran Duc Thao, *Origins of Language and Consciousness,* Boston Studies in the Philosophy of Science (ed. by Robert S. Cohen and Marx. W. Wartofsky), Volume LVI. 1979 (forthcoming).
119. Jerzy Pelč, *Semiotics in Poland, 1894 - 1969.* 1977, XXVI + 504 pp.
120. Ingmar Pörn, *Action Theory and Social Science. Some Formal Models.* 1977, X + 129 pp.
121. Joseph Margolis, *Persons and Minds, The Prospects of Nonreductive Materialism,* Boston Studies in the Philosophy of Science (ed. by Robert S. Cohen and Marx W. Wartofsky), Volume LVII. 1977, XIV + 282 pp. Also available as paperback.
122. Jaakko Hintikka, Ilkka Niiniluoto, and Esa Saarinen (eds.), *Essays on Mathematical and Philosophical Logic. Proceedings of the Fourth Scandinavian Logic Symposium and of the First Soviet-Finnish Logic Conference, Jyväskylä, Finland, 1976.* 1978, VIII + 458 pp. + index.
123. Theo A. F. Kuipers, *Studies in Inductive Probability and Rational Expectation.* 1978, XII + 145 pp.
124. Esa Saarinen, Risto Hilpinen, Ilkka Niiniluoto, and Merrill Provence Hintikka (eds.), *Essays in Honour of Jaakko Hintikka on the Occasion of His Fiftieth Birthday.* 1978, IX + 378 pp. + index.
125. Gerard Radnitzky and Gunnar Andersson (eds.), *Progress and Rationality in Science,* Boston Studies in the Philosophy of Science (ed. by Robert S. Cohen and Marx W. Wartofsky), Volume LVIII. 1978, X + 400 pp. + index. Also available as paperback.
126. Peter Mittelstaedt, *Quantum Logic.* 1978, IX + 149 pp.
127. Kenneth A. Bowen, *Model Theory for Modal Logic. Kripke Models for Modal Predicate Calculi.* 1978, X + 128 pp.
128. Howard Alexander Bursen, *Dismantling the Memory Machine. A Philosophical Investigation of Machine Theories of Memory.* 1978, XIII + 157 pp.
129. Marx W. Wartofsky, *Models: Representation and Scientific Understanding,* Boston Studies in the Philosophy of Science (ed. by Robert S. Cohen and Marx W. Wartofsky), Volume XLVIII. 1979 (forthcoming). Also available as a paperback.
130. Don Ihde, *Technics and Praxis. A Philosophy of Technology,* Boston Studies in

the Philosophy of Science (ed. by Robert S. Cohen and Marx W. Wartofsky), Volume XXIV. 1979 (forthcoming). Also available as a paperback.

131. Jerzy J. Wiatr (ed.), *Polish Essays in the Methodology of the Social Sciences,* Boston Studies in the Philosophy of Science (ed. by Robert S. Cohen and Marx W. Wartofsky), Volume XXIX. 1979 (forthcoming). Also available as a paperback.
132. Wesley C. Salmon (ed.), *Hans Reichenbach: Logical Empiricist.* 1979 (forthcoming).
133. R.-P. Horstmann and L. Krüger (eds.), *Transcendental Arguments and Science.* 1979 (forthcoming). Also available as a paperback.

SYNTHESE HISTORICAL LIBRARY

Texts and Studies
in the History of Logic and Philosophy

Editors:

1. M. T. Beonio-Brocchieri Fumagalli, *The Logic of Abelard.* Translated from the Italian. 1969, IX + 101 pp.
2. Gottfried Wilhelm Leibniz, *Philosophical Papers and Letters.* A selection translated and edited, with an introduction, by Leroy E. Loemker. 1969, XII + 736 pp.
3. Ernst Mally, *Logische Schriften,* ed. by Karl Wolf and Paul Weingartner. 1971, X + 340 pp.
4. Lewis White Beck (ed.), *Proceedings of the Third International Kant Congress.* 1972, XI + 718 pp.
5. Bernard Bolzano, *Theory of Science,* ed. by Jan Berg. 1973, XV + 398 pp.
6. J. M. E. Moravcsik (ed.), *Patterns in Plato's Thought. Papers Arising Out of the 1971 West Coast Greek Philosophy Conference.* 1973, VIII + 212 pp.
7. Nabil Shehaby, *The Propositional Logic of Avicenna: A Translation from al-Shifā: al-Qiyās,* with Introduction, Commentary and Glossary. 1973, XIII + 296 pp.
8. Desmond Paul Henry, *Commentary on De Grammatico: The Historical-Logical Dimensions of a Dialogue of St. Anselm's.* 1974, IX + 345 pp.
9. John Corcoran, *Ancient Logic and Its Modern Interpretations.* 1974, X + 208 pp.
10. E. M. Barth, *The Logic of the Articles in Traditional Philosophy.* 1974, XXVII + 533 pp.
11. Jaakko Hintikka, *Knowledge and the Known. Historical Perspectives in Epistemology.* 1974, XII + 243 pp.
12. E. J. Ashworth, *Language and Logic in the Post-Medieval Period.* 1974, XIII + 304 pp.
13. Aristotle, *The Nicomachean Ethics.* Translated with Commentaries and Glossary by Hypocrates G. Apostle. 1975, XXI + 372 pp.
14. R. M. Dancy, *Sense and Contradiction: A Study in Aristotle.* 1975, XII + 184 pp.
15. Wilbur Richard Knorr, *The Evolution of the Euclidean Elements. A Study of the Theory of Incommensurable Magnitudes and Its Significance for Early Greek Geometry.* 1975, IX + 374 pp.
16. Augustine, *De Dialectica.* Translated with Introduction and Notes by B. Darrell Jackson. 1975, XI + 151 pp.

17. Arpád Szabó, *The Beginnings of Greek Mathematics.* 1979 (forthcoming).

18. Rita Guerlac, *Juan Luis Vives Against the Pseudodialecticians. A Humanist Attack on Medieval Logic.* Texts, with translation, introduction and notes. 1978, xiv + 227 pp. + index.

SYNTHESE LANGUAGE LIBRARY

Texts and Studies
in Linguistics and Philosophy

Managing Editors:

JAAKKO HINTIKKA
Academy of Finland, Stanford University, and Florida State University (Tallahassee)

STANLEY PETERS
The University of Texas at Austin

Editors:

EMMON BACH (University of Massachusetts at Amherst)
JOAN BRESNAN (Massachusetts Institute of Technology)
JOHN LYONS (University of Sussex)
JULIUS M. E. MORAVCSIK (Stanford University)
PATRICK SUPPES (Stanford University)
DANA SCOTT (Oxford University)

1. Henry Hiż (ed.), *Questions.* 1977, xvii + 366 pp.
2. William S. Cooper, *Foundations of Logico-Linguistics. A Unified Theory of Information, Language, and Logic.* 1978, xvi + 249 pp.
3. Avishai Margalit (ed.), *Meaning and Use. Papers Presented at the Second Jerusalem Philosophical Encounter, April 1976.* 1978 (forthcoming).
4. F. Guenthner and S. J. Schmidt (eds.), *Formal Semantics and Pragmatics for Natural Languages.* 1978, viii + 374 pp. + index.
5. Esa Saarinen (ed.), *Game-Theoretical Semantics.* 1978, xiv + 379 pp. + index.
6. F. J. Pelletier (ed.), *Mass Terms: Some Philosophical Problems.* 1978, xiv + 300 pp. + index.

RAY OND